Umweltsoziologie

Edited by

Prof. Dr. Matthias Groß, Helmholtz-Zentrum für Umweltforschung, Leipzig und Friedrich-Schiller-Universität Jena
Prof. Dr. Stefanie Hiß, Friedrich-Schiller-Universität Jena
Prof. Dr. Harald Heinrichs, Leuphana Universität Lüneburg
Prof. Dr. Ortwin Renn, Institut für transformative Nachhaltigkeitsforschung e. V. (IASS), Potsdam

Volume 5

Laura Spengler

Sufficiency as Policy

Necessity, Possibilities and Limitations

Nomos

The Deutsche Nationalbibliothek lists this publication in the Deutsche Nationalbibliografie; detailed bibliographic data are available on the Internet at http://dnb.d-nb.de

a.t.: Lüneburg, Universität, Dissertation, 2017

Original title: Sufficiency as Policy: A Theoretical Analysis of the Necessity, Possibilities and Limitations of a Political Implementation of the Sufficiency Strategy

ISBN 978-3-8487-4205-9 (Print)
978-3-8452-8474-3 (ePDF)

British Library Cataloguing-in-Publication Data
A catalogue record for this book is available from the British Library.

ISBN 978-3-8487-4205-9 (Print)
978-3-8452-8474-3 (ePDF)

Library of Congress Cataloging-in-Publication Data
Spengler, Laura
Sufficiency as Policy
Necessity, Possibilities and Limitations
Laura Spengler
327 p.
Includes bibliographic references.

ISBN 978-3-8487-4205-9 (Print)
978-3-8452-8474-3 (ePDF)

1st Edition 2018

Acknowledgements

While this thesis is genuinely my own work and any mistakes or inconsistencies are my own fault, it has clearly benefitted from other people's help and influence.

First of all, I thank my supervisors Harald Heinrichs and Thomas Schramme for their many helpful comments on earlier drafts and helpful discussions during the work on this thesis. Thank you for your ideas, time and support, and for always answering my questions promptly. I also thank the members of their respective PhD colloquia at Lüneburg and Hamburg Universities for their helpful comments on particular chapters and aspects of my thesis. Further, I thank my third thesis examiner Thomas Schomerus for accepting this task and for the good exchange on diverse issues in environmental policy and law we have had over the past few years.

I also thank my colleagues and ex-colleagues at Ökopol, in particular Dirk Jepsen and Norbert Reintjes, who have supported my work on this thesis significantly over the years by allowing me part-time work and not overloading my work capacities with too many projects, and who have listened to and commented on presentations of interim versions of my research project.

I owe Lisa Rödig and Felix Schwabedal special thanks, who contributed with helpful comments on a full version of this thesis. I also thank all the others with whom I have discussed the diverse related issues and who have in one way or another influenced my work. I would like to mention here in particular Christian Baatz and Julian Müller.

Finally, I thank my family for their patience and support, and for tolerating my work on this thesis over the years. Thank you Christian, also for our many helpful and interesting discussions, and thank you little Bruno for not causing me more sleepless nights.

Overview of contents

Detailed List of Contents

List of tables and boxes

1. Introduction

Can we solve the most pressing environmental problems through the application of modern science and technology, or do we need changes in lifestyles and consumption levels instead, as technology is a source of the existing problems rather than a solution to these? The tradition of debating this question within the environmental, and later, the sustainability discourse is more or less as old as the concern for the environment itself. More recently, however, agreement with a middle course between these two positions, rejecting both overly optimistic technological visions on the one hand as well as overly naïve "back to nature" and "simple living" scenarios on the other, seems to have increased. According to this position, both technology and changes in consumption patterns are needed in order to mitigate the most pressing environmental problems. Empirical data on climate change and other environmental issues gives strong evidence that typical lifestyles in rich countries are neither applicable to the entire world nor maintainable in the long term. Thus, something like "sufficiency"[1], that is, decreasing the levels of consumption associated with significant environmental pressures in these countries, is needed besides technical improvements in order to avoid, or at least mitigate, harmful environmental change.

However, this subject is very unpopular within political discussion. Policies and political programmes aiming at sustainability are very much focused on technical solutions well-suited to fitting into the prevalent paradigm of increasing rather than redistributing the cake of wealth and material assets. At the same time, at least among environmental activists and academics, the debate has moved forward: For many, the question of whether people in wealthy countries need to consume less is now clearly to be answered in the affirmative; thus, they have started to ask whether action in this respect should not only be taken by individuals, but also on the policy level. Sufficiency has long been understood exclusively as an

1 Many discussions revolve around the term sufficiency (cf. section 3.2) and the advantages and disadvantages of its use. I do not insist on this term, but I use it because it is established in sustainability discourse (especially in the part of it conducted in German language).

individualistic and voluntary strategy as it seemed too daring and not possible within liberal societies to attempt to influence what and how much individuals consume.

Indeed, such attempts can be dangerous for politicians: A rather incidental proposal on page 164 of the Green Party's programme for the elections to the German *Bundestag* in 2013, consisting of two short and sober sentences, triggered a storm of protest in the media and public discussion in advance of the elections and potentially cost the party a few percentage points of votes. In the context of a description of the practices and adverse effects of factory farming and high rates of meat consumption, the programme stated (Bündnis 90/Die Grünen 2013): "Public canteens will have a pioneer function. Offers of vegetarian and vegan dishes as well as a 'veggie day' should become the standard."[2] Even if this proposal sounds rather harmless—after all, the intention was not to establish a mandatory vegetarian day, nor is anybody obliged to have lunch in a canteen that day—the more evil element to it was what some newspapers and representatives of rival parties made of it. The Greens were accused of aiming to prohibit eating meat entirely, of expressing an "arrogant prohibitor mentality" and "totalitarian traits", and were assumed to be striving for an "educational dictatorship" (Focus Online 2013, Spiegel Online 2013, Stern 2013). Even if due to the approaching elections this debate seemed to be a campaign by their opponents against the Greens, the episode nevertheless demonstrates that much care is needed in raising the idea of influencing individual consumption levels by means of political measures.

However, it is now increasingly recognised that individual consumption choices have always been influenced by certain framework conditions determined by society, including policy level. Sometimes, individuals are simply not in a position to adopt, for instance, an environmentally sound mobility pattern, because the necessary infrastructure is not supportive of that aim, or (e.g. economic) incentives exist to make certain unsustainable consumption decisions. In addition, certain consumption patterns may have potential harmful consequences that will affect others living elsewhere on the planet or in the future. Even in a liberal society, there must be ways to restrict individual liberty if individual action is likely to have negative effects on other people. However, research on the reasons for and need to adopt sufficiency on a political level is very limited, and most of

2 My translation.

the available publications on sufficiency as policy focus very much on its practical application and implementation, while omitting the step of justifying this in detail.

Aim and framework of this study

This study thus explores the question of whether it is justified, or even necessary, within wealthy liberal-democratic societies to adopt sufficiency also as a political strategy in order to strive for sustainability. On this basis, it further discusses possible environmental policies in this sense from a more practical perspective, as well as any limitations to them. The aim is, ultimately, to develop a conception[3] of "sufficiency as policy" based on the discussions and framework laid out in this study's preceding chapters. An important clarification in this context concerns the terms "political", "political level" or "policy". Of course, there are many actors who have long argued in favour of sufficiency as a necessary strategy for sustainability and who are, depending on the use of the term, political actors, for instance environmental organisations. My use of the term here is, however, more restricted and includes only activities by the state that aim at collective organisation. The reasons for and details of this focus will be explained at the beginning of chapter 4, before I enter into a discussion of sufficiency as an individual and as a political approach.

The basic framework chosen for the analysis is, as mentioned above, liberal democracy. At the same time, this means that other state systems are excluded from the conclusions and proposals that will be made. The reason for this choice is simply that in most wealthy societies, which are the target of the sufficiency debate, liberal democracy is the system we currently have, which many people value and which can be expected to be here to stay for quite some time. In addition, the conception to be developed should be pragmatic and acceptable to as many people as possible

---

3 John Rawls has distinguished between the terms "concept" and "conception" in the context of justice theory in the following way: the concept of justice describes the basic role of the idea as such, so there is only one concept of justice; different people may hold different conceptions of justice which flesh out the basic idea with specific principles they subscribe to (Rawls 1999 a: 5). In this sense, it is useful to speak of a specific "conception of sufficiency as policy", the principles of which are outlined in section 5.3.

in order to be feasible, which is why it makes sense not to propose something based on a different political or economic system or embedded into a different value system.

Further, the concern with sufficiency has developed within sustainability discourse; thus the following analysis has to be embedded into this discourse. As interpretations of sustainability itself vary significantly, it is necessary to specify in detail the conception of sustainability on which the following discussion is based. Sustainability in this sense is taken as a desirable state liberal-democratic societies should (at least) strive for, as it aims to preserve their livelihood. It can thus be summarised that two components form the theoretical framework of this study: liberal democracy as the given framework conditions, and sustainability as one aim development should strive for. Further details of the adopted conception of sustainability will be provided later, but two things should already be said here. First, the perspective taken is primarily an anthropocentric one. It relies on arguments related to the preservation of the livelihood of human beings and the requirement of avoiding harm to them, and does not consider in detail arguments related to the rights or the potential "intrinsic value" of other elements of nature—again, mainly for the purpose of making the argumentation accessible to as many people as possible. Second, the focus here lies on the environmental aspects of sustainability. Sustainability is often said to be a concept that requires to balance demands from different fields[4] and does not only incorporate a concern for the environment. It is, however, the environmental concern that this study deals with, as the related problems are already falling or will likely fall back on human beings in the future and severely affect their health and livelihood. Thus, the other aspects or dimensions of sustainabiliy will be dealt with only as far as they depend on the integrity of the environment. It will not be discussed in detail what, for example, social demands that exist independently of environmental issues may be contained in the vision of sustainability in addition.

This focus also corresponds to my personal research focus: as an environmental scientist, I obviously deal with environmental issues, and in addition I am used to combining the perspectives of different disciplines on a certain research subject. The issue of sufficiency and policy touches upon a variety of topics that are often dealt with separately within a number of

---

4 On this issue, cf. section 2.3.1.

different disciplines. This study assembles strands of debate mainly from political philosophy, environmental ethics, policy analysis and the (already interdisciplinary) field of sustainability science, as well as insights from economics and sociology. It is therefore an interdisciplinary piece of research and draws on the systematic review of an extensive base of literature.

Further, I analyse my subject from a theoretical perspective and the following discussion does not contain own empirical research components. However, the analysis builds on empirical data about environmental issues. In order to set the scene and summarise some of the currently most significant problems, several examples and according data will be provided on the next few pages before I go on to present the theoretical framework.

Status quo regarding selected environmental issues and their potential harmful consequences

I will briefly present a selection of environmental problems and their (potential) harmful consequences in the following, for which the current trend is not sustainable and technical solutions alone do not seem to be a sufficiently effective answer: anthropogenic greenhouse gas emissions and climate change, the overuse of natural resources and the loss of biodiversity. At the same time, these are some of the major global environmental issues, which have the potential of impacting Earth system functions and thus entail significant risks for human beings (cf. Steffen et al. 2004, 2015, Rockström et al. 2009 a, 2009 b). Even if there have been promising developments lately—for instance, the Paris Agreement on climate change (UNFCCC 2015) or the agreement on Sustainable Development Goals (UN 2015 b)[5]—greater efforts are required to get these problems under control, most likely including solutions in the sense of sufficiency. The issues presented here are examples only and do not constitute a conclusive list; the intention is just to show the urgency of the subject of this study.

Before turning to the environmental problems mentioned, I would like to mention two examples of environmental issues for which technical so-

5 It should, however, be noted that unlike the Paris Agreement, the Sustainable Development Goals are mostly rather global and tentative.

lutions have proven quite successful: first, the development with regard to a number of air pollutants, and second, the destruction of the stratospheric ozone layer. Air pollution varies greatly among regions worldwide (cf. UNEP 2012: 41 ff.), but as environmental trends in most industrialised countries show, efforts over the past 30 to 40 years have been quite successful in significantly reducing the major pollutants, including sulphur dioxide, nitrous oxides, particulate matter, non-methane volatile organic compounds, and ammonia (cf. e.g. EEA 2012 a: 60 ff.)[6]. Sulphur dioxide emissions, for instance, decreased by between 55 and 93% in the major Western European economies between 1980 and 2004 and by 73% in Europe as a whole, with the main reasons for this being the implementation of abatement technologies and a turn towards fuels with lower sulphur content, and not because of a decline in the demand for sulphur producing activities such as energy and heat production (Verstreng et al. 2007).

As regards the ozone layer, the existing international agreements have been very successful in reducing ozone-depleting substances, as technical substitutes for these substances were available. To be sure, some of the problematic effects will still remain for another 10 to 35 years (depending on the region), as the substances under concern have very long atmospheric lifetimes and some chlorofluorocarbon is still released, for example from inadequate disposal of old refrigeration appliances. However, the cause of the problem has been abated effectively: the use of ozone depleting substances decreased dramatically, expressed in ozone depleting potential, by around 97% in 2010 compared to the level in the late 1980 s (cf. UNEP 2012: 51 ff.).

However, technical solutions appear to be less effective and more costly for other environmental problems, some of which will be presented on the following pages.

6 Which does not mean that the emissions have reached harmless levels for all pollutants yet. In addition, it should be noted that it is subject to debate to what extent the shift of polluting industries to other countries with less stringent environmental standards has also contributed to the decrease in pollutant emissions in wealthy countries (cf. e.g. Peters/Hertwich 2006, EEA 2014: 31 ff.).

Greenhouse gas emissions and climate change

Anthropogenic greenhouse gas emissions and related climate change pose one of the most pressing environmental problems humanity currently faces. Climate change might not only bring about significant changes to ecosystems and the environment around the world, but these also pose enormous risks to humans. To summarise the status quo of climate research briefly (based on IPCC 2014[7]):

The last three decades have each been successively warmer than any preceding decade since the beginning of temperature recording. The globally averaged combined land and ocean surface temperature increased by around 0.85°C over the period from 1880 to 2012. Anthropogenic greenhouse gas emissions are currently higher than ever, with nearly 50 billion tonnes in 2010, and are, together with other anthropogenic drivers, "extremely likely" to have been the dominant cause of the observed warming (IPCC 2014: 4). The still "safe" share of carbon dioxide in the atmosphere is estimated to be 350 parts per million, while the current level is already at 398.5 parts per million (Steffen et al. 2015)[8]. Various changes within the climate system have been observed, including the following: ocean warming, changes to precipitation amounts differing among regions (increasing in some while likely decreasing in others), changes to ocean surface salinity due to higher evaporation in some and higher precipitation in other areas, ocean acidification amounting to a 26% increase in acidity since beginning of the industrial era due to oceanic uptake of carbon dioxide, shrinkage of polar ice sheets and glaciers worldwide (also differing regionally—in some places in Antarctica ice mass seems to have increased), and a global rise in the sea level amounting to around 0.19 metres on average between 1901 and 2010.

These changes to the climate system have the following impacts: changes to the regional availability of water resources, geographic shifts

---

7 The IPCC report very carefully identifies the level of confidence of scientific findings mentioned in the report, from very low to very high (cf. IPCC 2014: 2), and, in addition, characterises these in some cases with a quantified likelihood. In order to improve readability, I do not repeat these confidence levels and probabilities in each case here. The observed changes and impacts cited here are characterised by at least medium to very high confidence levels.

8 For such "safe" thresholds or "planetary boundaries", a concept developed by Rockström et al. (2009 a, 2009 b) and Steffen et al. (2015), see section 2.3.4.

and differences in the abundance of species, (mainly negative) impacts on crop yields, higher frequency of heat waves, an increase in heat-related and a decrease in cold-related human mortality, and more frequent extreme sea levels as well as more frequent extreme precipitation events, implying greater risks of flooding. If greenhouse gases are emitted into the atmosphere at a continuously high level in the future, these impacts will, along with temperatures rising by between 2 and 4.8°C until the end of the century, most likely be aggravated, and, depending on the extent of global warming, lead to long-lasting and irreversible changes that will affect human beings and ecosystems. The risks involved are distributed unevenly across the globe and "are generally greater for disadvantaged people and communities in countries at all levels of development" (IPCC 2014: 13). Besides directly weather-related risks, climate change in the future might undermine food security with regard to fisheries and crops, lead to a further reduction in water resources in dry subtropical regions, and impact on human health through the extension of the infection area of some diseases. In a scenario of high temperature increase, in some areas of the world even common human activities might be compromised, such as working outdoors. All in all, the impacts of climate change are projected to make poverty reduction more difficult and increase the displacement of people.

Except in a scenario of zero or even net negative emissions by the year 2100, warming is expected to continue after 2100, and temperatures will remain more or less constant at the elevated level already reached by then even after completely anthropogenic carbon dioxide emissions have stopped. This would, however, not imply stabilisation for all other parts of the climate system, as the interactions within this system are very slow. The effects of anthropogenic greenhouse gas emissions and global warming are therefore expected to last hundreds to thousands of years after the stabilisation of the mean surface temperature.

To curb climate change and limit the related risks, substantial reductions in greenhouse gas emissions ("mitigation") are necessary. In addition, adaptation measures can be adopted to further decrease the risks; however, their effectiveness is expected to be limited in particular in case of greater rates of temperature rises. Efforts to reduce greenhouse gas emissions are required in particular in wealthy countries, whose emissions are relatively high, even if emerging economies are catching up. For instance, emissions in the EU accounted for about 13% of worldwide greenhouse gas emissions, while it is inhabited by only about 7% of the world's population, and emissions in the US accounted for nearly 19% of the total, while it is inhabited by 4.5% of the world's population[9]. Per-capita emissions in these countries are significantly above the level that would be sustainable on average[10]. These figures do not yet take into account the net emissions "embedded" in goods imported to rich countries, which are also significant (cf. Peters/Hertwich 2006, Peters et al. 2011, Wiebe et al. 2012, UNEP 2012: 21). Around 24% of greenhouse gas emissions caused by consumption in the EU occurred outside EU borders in 2008 (EEA 2014: 34).

The problem of greenhouse gas emissions and global warming has been recognised as a significant environmental problem for decades now and has been dealt with on policy level since the 1970 s. At least since the adoption of the United Nations Framework Convention on Climate Change (UNFCCC) in 1992, political activity in this field has been intensive and countries worldwide have adopted countless measures aimed at decreasing greenhouse gas emissions. Typical measures are to increase the share of renewable energy technologies as a replacement for fossil fuels, and to increase the energy efficiency of industry plants and electrical appliances. Further cutting the emissions using technical measures is becoming increasingly difficult and cost-intensive in most industrialised countries. Recently, the parties to the UNFCCC agreed in Paris to keep the increase in the average global temperature "well below 2°C" above pre-industrial levels, and even to pursue efforts to limit the increase to 1.5°C (UNFCCC 2015, article 2). While this is a promising step, it remains to be seen what concrete action will be taken by the individual parties and whether they will be successful in achieving their respective emission reductions.

9 Based on data from http://unfccc.int/ghg_data/ghg_data_unfccc/time_series_annex_i/items/3841.php (last accessed August 2015).

10 Note that the statement that especially wealthy countries should reduce their emissions seems to silently endorse the idea that equal per-capita emissions would be just. While this is being discussed controversially (cf. sections 2.3.3, 5.2.2), I think it is clear that the emissions of wealthy countries are too high in any case, also if judged by most other distribution principles.

Use of natural resources

A further issue that has been receiving increasing attention over the last few years[11] is the use of natural resources. The term "natural resources" is used in various definitions in the literature on this subject, from very wide to very narrow ones. Some definitions include virtually every component of the environment: raw materials, environmental media (air, water, soil), flow resources (renewable energies), space, and genetic resources (cf. EC 2005, UBA 2012)[12]. This therefore includes assets that are used as "sources" of production as well as "sinks" that absorb emissions and waste, including the atmosphere as a sink for carbon dioxide. In this sense, the term resources seems to be more or less equivalent to the environment and its usefulness is therefore limited. Other definitions focus more on the source function[13] and the extraction of renewable and non-renewable materials, including for the purpose of nutrition, and including the use of land area (cf. e.g. EPA Network 2006: 2). In the following, this narrow understanding of the term is used.

Because of the diversity of the object of concern, determining resource use is more complicated than quantifying greenhouse gas emissions. Various indicators at different levels of aggregation have been proposed. Estimates exist, for instance, for the sum of all raw materials extracted globally. Included in this figure are biomass, fossil energy carriers, ores and industrial minerals, and construction minerals. In the early 21st century, raw material extraction ranged from 47 to 59 billion tons per year globally, while it amounted to only about 7 billion tons in 1900. Total raw material extraction is thus estimated to have increased by a factor of 8 from 1900 to 2005, and continues to rise at high rates (Krausmann et al. 2009, UNEP

11 For instance, in the context of the International Resource Panel launched by UNEP in 2007 (see http://www.unep.org/resourcepanel/, last accessed July 2015) and the Sustainable Development Goals, especially goals 6, 14, and 15 (UN 2015 b), and within the EU by means of numerous initiatives, a flagship initiative on resource efficiency (see http://ec.europa.eu/resource-efficient-europe/, last accessed July 2015), and the adoption of an action plan for the "Circular Economy" (EC 2015).

12 The Ecological Footprint is a possible (highly aggregated) indicator of such a rather wide understanding of resources, as it includes carbon emissions as the most important component (cf. e.g. WWF 2014 a).

13 Note that some of the components providing sink functions can also play an important role as sources, for instance water and soil. These therefore qualify as resources in a narrower sense as well.

2011 a: 10, 17)[14]. In a business-as-usual scenario—with industrial countries keeping their raw material use at the same levels and developing countries catching up to this level by 2050—,annual global material extraction would more than triple by mid-century (UNEP 2011 a: 28 f.). Resource uses that this figure does not include are, for instance, water use and land use[15]. The share of available freshwater resources used by agriculture, industry and municipalities shows strong regional differences. While amounting to only around 7% in developed regions, many less privileged regions experience water stress or even severe water scarcity, which means that between 25% and 80% of the renewable freshwater is extracted. Water scarcity[16] currently affects more than 40% of the world's population, and this share is projected to increase (UN 2015 a: 55). Land use can be quantified by various indicators, for instance as changes to the global forest cover, the area used to grow crops, or the soil area sealed for human dwellings and infrastructure. Even if the net loss of forests has slowed recently, from 8.3 million hectares annually in the 1990 s to 5.2 million hectares per year between 2000 and 2010, this still means that deforestation continues at significant rates (UN 2015 a: 52). At 62% the current global area of forested land, expressed as a share of the original forest cover, has fallen below the estimated "safe boundary" of 75% (Steffen et al. 2015). The main cause of deforestation is the worldwide expansion of cropland, which currently accounts for about 10% of the world land area[17] (UNEP 2014 b: 25 ff.). The land used to grow crops increased by 11% globally between the 1960 s and 2007, and is predicted to increase by another 8 to 33% by 2050 under business-as-usual conditions (UNEP 2014 b: 13). Sealed soil covered by settlements and infrastructure accounted for between 1 to 3% of the total land area in 2005, and is expected to increase to 4 to 5% by 2050 without policy interventions (UNEP 2014 b: 24).

---

14 As Krausmann et al. (2009: 2703) emphasise, in the case of such highly aggregated material use indicators, there is, of course, no simple linear relation between indicator value and environmental degradation, but still the indicators can be used as a proxy for environmental pressures resulting from human material extraction.

15 These are affected only indirectly through the extraction processes of the raw materials mentioned, see below.

16 Water scarcity, according to the UN definition, starts when more than 60% of renewable freshwater is extracted (see UN 2015 a: 55).

17 Note that deserts, glaciers and other extreme habitats account for approximately one third of the global land area.

The observed increase in the extraction of natural resources has resulted in a significant increase in environmental pressures, which exacerbate environmental quality for human beings and are (potentially) harmful. Environmental impacts are caused directly by the processes and technologies used for material extraction and by the fact that these resources are not available anymore in the place they have been removed from, as well as more indirectly by their further processing, use and disposal. These indirect impacts are not focused on so much here, as they depend on what exactly happens to the materials after extraction and not on the extraction itself. Direct impacts from biomass extraction include deforestation, soil degradation, water use and contamination, the impacts of fertilizer use, and biodiversity loss. As regards fish stocks, for example, marine fisheries have been declining since 1996[18] and 28.8% of stocks were overfished in 2011, and the share of overfished stocks is tending to increase (FAO 2014: 37). The extraction of fossil fuels and industrial minerals and ores generates severe land degradation at the extraction sites and casues the release of toxic substances in the mining process, and often requires high energy inputs. These impacts will become even more relevant with the exploitation of "unconventional" fuels such as shale gas. With regard to construction minerals, besides land degradation, their impacts are mostly indirect, including those caused by transportation and further processing, such as energy intensive cement production (Krausmann et al. 2009: 2703, UNEP 2011 a: 25 f.). The negative consequences of freshwater scarcity and decreasing water quality for local ecosystems and human beings are quite obvious, as this resource is a prerequisite for all forms of life on Earth. Significant changes in land use, such as the extension of cropland, decrease biodiversity, influence the climate system and the hydrological cycle, and jeopardise the regulatory capacity of these systems (Rockström et al. 2009 a, EEA 2015: 59 f.). Even if the intensification and extension of agriculture also helps to feed the growing world population, 1.6 billion people directly depend on forests for their livelihoods, according to the UN (2015: 52). The growing demand for food, feed and recently also energy crops is one of the factors that make food prices more volatile, which compromises food security especially for the poor (UNEP 2014 b: 37 ff.). Food prices, along with prices of other resources such as raw materials

18 At the same time, global fish production is increasing due to a steady increase in aquaculture and inland capture production. Per-capita fish consumption is also increasing, despite ongoing population growth (FAO 2014).

and metals, have been falling in the long term for decades until the late 1990 s, and have since then started to rise again steeply, except for a collapse in the curve during the financial crisis around 2008 (UNEP 2014 a: 23).

The steep increase in material use after 1950 is, besides population growth[19], mainly due to further industrialisation in Europe, the USA, Japan, and the USSR. Further growth in material extraction since the year 2000 can mainly be attributed to emerging economies such as China, India and Latin American countries. However, fully industrialised countries, home to 15% of the world's population, are still dominant in global materials use, using one third of materials in the year 2000, and this is without considering the indirect use of materials embedded in imports (Krausmann et al. 2009: 2703). As for greenhouse gases, the resource use of wealthy countries is also significantly higher if the resource input from outside needed for consumption within these countries is included (Peters/Hertwich 2006, Steen-Olsen et al. 2012, Wiebe et al. 2012). For instance, 56% of the land footprint associated with EU consumption is located outside EU borders (EEA 2015: 41). According to the United Nations Environment Programme, it will not be possible to eradicate poverty and meet the needs of nine billion people by 2050 without absolute resource use reductions in developed economies (UNEP 2011 a: 32). Thus, even if the use of resources and materials has become more prominent on political agendas recently, more efforts are required to bring resource use in line with the demands of sustainability.

Biodiversity loss

A third and final example is the loss of biodiversity. The term biological diversity, according to the *Convention on Biological Diversity*,

> *"means the variability among living organisms from all sources including, inter alia, terrestrial, marine and other aquatic ecosystems and the ecological complexes of which they are part; this includes diversity within species, between species and of ecosystems."* (UN 1992, article 2)

19 While global material extraction multiplied eightfold between 1900 and the early 2000 s, material use per capita merely doubled as the world population increased by a factor of four (UNEP 2011 a: 10).

Its loss has always been a concern of conservation groups because it is as such regrettable if a species is lost or if a complex of ecosystems becomes less diverse, but in more recent times in addition, concerns have been growing that biodiversity loss may have also more direct adverse effects on human beings.

Biodiversity is being lost at dramatic rates: measured by the number of species becoming extinct per million species-years, the current extinction rate of more than 100 exceeds the pre-industrial rate by its 100 to 1000-fold level (Rockström et al. 2009 a). There are manifold other methods and indicators to measure the development of biodiversity, such as the Red List Index based on the IUCN's Red List of Threatened Species and the Living Planet Index by the WWF. These do not only include the number of species extinctions but also, for instance, the number of threatened species within a certain group (mammals, birds, corals etc.), their distribution, and/or the number of populations. All these show the same declining trend and indicate that biodiversity is decreasing at the levels of populations, species, as well as habitats (cf. UNEP 2012: 145, WWF 2014 a: 16, UN 2015 a: 57).

There are five major reasons for biodiversity loss: Habitat loss and degradation due to land use change, climate change, overexploitation of natural resources such as fisheries, pressure due to invasive alien species, and pollution, in particular from nitrogen and phosphorus (Millennium Ecosystem Assessment 2005: 16, UNEP 2012: 134, EEA 2012 a: 36, 2015: 56). These drivers vary in importance from place to place, but overall they continue to increase the pressure on biodiversity, in particular as they often interact and reinforce each other (EEA 2012 a: 36).

Compared to the two problems described above, climate change and increasing resource use, which are also among the drivers of biodiversity loss, it is more difficult to see how the loss of biodiversity itself might harm human beings. The loss of a particular species or ecologically valuable site will most likely have no perceivable impact on human welfare at all (as said above, apart from the regrettable fact that it is not there any longer), but scientists emphasise that it is the scale at which biodiversity is currently being lost which also entails significant risks for humanity, as it may disturb the Earth system (Rockström et al. 2009 a, Steffen et al. 2015). The loss of biodiversity increases the vulnerability of ecosystems and makes them less resilient to other influences, with the result that the functions they have for human beings, including sink functions and providing resources, are endangered. For example, if a whole functional group of species such as pollinators decreases significantly within an ecosystem, the system is no longer able to provide the services it previously did and may in addition change irreversibly into a different state (Folke et al. 2004). Genetic diversity as such also makes ecosystems more resilient and provides them with the capacity to adapt to external changes (Steffen et al. 2004: 33 ff., 2015). Further, studies indicate that biodiversity loss increases the transmission of infectious diseases among humans, animals as well as plants (Keesing et al. 2010).

There is already an increasing number of policy responses to the loss of biodiversity, such as the establishment of protected areas and action to manage invasive alien species. However, these have so far failed to reduce the pace at which biodiversity is being lost (UNEP 2012: 134). It seems that local action to preserve certain areas and prevent alien species from invading is not sufficient to deal with the problem, as the drivers of biodiversity loss are very diverse and operate to a large degree on a global and systemic scale. Therefore, reducing the pace at which biodiversity is lost appears to require a reduction in the general anthropogenic burden on the environmental carrying capacity. While technical solutions can play a role in this (Brook/Bradshaw 2012), it is unlikely that they can deal effectively with climate change, overexploitation, habitat loss, nitrogen and phosphorus pollution, and invasive species simultaneously (cf. section 3.1.4).

The exemplary environmental issues described here are serious and require stronger efforts in order for us to achieve a sustainable development. This study argues for a response in this sense that has so far been widely neglected.

Structure of the study

The argumentation presented in this study is, besides this introduction and the concluding chapter, structured in four major chapters. Chapter 2 comprises the theory building blocks on which the analysis that follows is based—liberal democracy and sustainability—and clarifies what characterises these as used here. Chapter 3 describes and analyses different "sustainability strategies", contrasting technical strategies with sufficiency, and summarises their respective limitations. Chapter 4 is the most comprehensive one and consists of four parts: it first analyses why individual sufficiency strategies are not sufficiently effective and why it is suggested that this be dealt with on the political level, too; second, it provides the rationale for doing this within a liberal state; third, it discusses some special difficulties that arise if one intends to interfere with consumption levels; and finally, it anticipates a number of critical perspectives that could be raised with regard to the approach suggested here and tries to give answers to these. Most of chapter 5 is less abstract than the preceding chapters and stands more in the tradition of practical policy analysis. It discusses potential "sufficiency policy" instruments, including three policy examples—environmental taxes, personal carbon trading, and planning instruments for urban infrastructure to reduce motorised individual transport. It further presents a conception of "sufficiency as policy" that contains the essence of what has been elaborated in the previous chapters. Chapter 6, then, contains a summary and some concluding remarks.

2. Theoretical framework: liberal democracy and sustainability

Introduction

This chapter aims to lay out the theoretical framework which prepares the ground for the analysis in the later chapters. It proceeds from the more general to the more specific: starting with a presentation of the most relevant lines of debate with regard to liberalism and democracy, it also analyses what keeps both components, the liberal and the democratic one, together. As stated in chapter 1, I take the liberal-democratic society as a given, and thus there is no need for arguments on why this type of society may be particularly valuable or superior to other systems[20]. Afterwards, the chapter provides a somewhat more comprehensive presentation of the most important lines of debate within the sustainability discourse. This comprises, most notably, whether an anthropocentric view should be adopted or not, what conception of justice is most appropriate in the context here, what can be learned from the debate on "weak" versus "strong" sustainability, and what can be said about the relationship between sustainability and development or economic growth.

The aim is to develop an understanding of what constitutes a liberal democracy, on the one hand, and of sustainability on the other, *for the purposes of this study*. The last half sentence highlights an important point: the task of this chapter is mainly to provide a theoretical basis to be able to derive the arguments related to sufficiency and policy later on, and it should therefore not be seen as exhaustive in documenting all theory debates surrounding liberalism, democracy, or sustainability. The sustainable liberal democracy presented here does not necessarily represent my personally favoured version either, but rather a robust version that contains certain minimal requirements which allow to achieve a rather broad agreement with various theory strands.

---

20 Note that much philosophical literature, for instance that on democracy, is dedicated to the analysis of instrumental and intrinsic reasons that speak in favour of this state form (cf. e.g. many of the contributions in Christiano 2003).

2.1 Liberalism

In this chapter, I will discuss those aspects related to liberalism which are most relevant to my subject here, as it is not possible to give a full picture of all the debates surrounding liberalism on just a couple of pages.[21] I therefore will also not provide any historical overview and not cite all the classical thinkers.

There is not a single, clear concept of "liberalism"[22]. As Gaus and Courtland (2010) point out: "Given that liberalism fractures on so many issues—the nature of liberty, the place of property and democracy in a just society, the comprehensiveness and the reach of the liberal ideal—one might wonder whether there is any point in talking of 'liberalism' at all." Nonetheless, they continue: "It is not, though, an unimportant or trivial thing that all these theories take liberty to be the grounding political value."

This hints at the presumption in favour of liberty, which plays a key role in most theories of liberalism. It states that in cases where liberty is to be restricted, the burden of proof is on those who want to restrict it, so the restriction needs to be justified by other—weightier—reasons (Feinberg 1984: 9).[23]

2.1.1 Liberty

What exactly liberty means is, as indicated by Gaus and Courtland, controversial. Or more explicitly, in the words of the Concise Oxford Dictio-

21 Additional aspects that are frequently discussed within liberal theorising and are not comprehensively covered here include at least: the primacy of the individual as opposed to the role of groups, meaning that the individual is "the fundamental element of analysis in political theorising" (Talisse 2005: 17, cf. also Raz 1986: 17 f., Avnon/de-Shalit 1999: 5), the role of autonomy within liberalism (Christman/Anderson 2005, Gaus 2005, Talisse 2005: 16 ff.), as well as theories of justice, which are dealt with here in a later chapter (see section 2.3.3).

22 One view on liberalism I do not cover here is the rather pejorative meaning that appeared in the USA and also in Britain in the late 20th century. In the US, the term "liberalism" is associated with rather leftist welfare state policies, while in the traditional European understanding the focus is—quite contrarily—on the limitation of state interference, as will be explained below.

23 Cf. Raz' critical comments on the presumption of liberty (1986: 8 ff.).

nary of Politics (McLean/McMillan 2009) in its entry on "freedom"[24]: "Unfortunately, and as a reflection of the importance which has been attached to ideas of freedom, almost every aspect of its characterization is controversial."

Historically, there are the two interpretations: "negative" and "positive" liberty, most famously framed by Isaiah Berlin (1969). While negative liberty refers rather plainly to a situation in which someone is free from external restrictions[25], particularly from state coercion, the concept of positive liberty is more comprehensive. It places more emphasis on the ability of someone to actually take control of his or her life and act autonomously. As Berlin insisted in his warning against positive conceptions of liberty in view of the totalitarian regimes experienced until the middle of the 20th century, the two interpretations are not just simply two sides of the same coin but should be seen as incompatible. Theorists who favour a conception of positive liberty have been accused of promoting certain ideals of human flourishing, for which negative liberties must be sacrificed, and therefore of disregarding the risk of drifting towards tyranny. Negative liberty theorists, on the other hand, may be accused of disregarding the importance of the availability of options to act freely, for example the availability of material resources or education. Endorsing a negative conception of freedom, however, does not imply that it must always come first and trump all other values, for instance equality or security (Kelly 2005: 54).

To overcome the "artificial" dichotomy of negative and positive liberty, MacCallum (1967) suggested conceiving of freedom as a single, unified concept. Within this, freedom can always be seen as a "triadic" relation of the following form: an agent is free (or not free) from something to do (or not do) something. Upon this basic concept, according to MacCallum, manifold interpretations of freedom are possible. Nonetheless, the positive and negative notions are still the most influential concepts in discussions of liberty. Besides these two, other "third" and "fourth" concepts have been developed since, and the discussions are still ongoing (cf. Spector 2010).

---

24 The terms "freedom" and "liberty" are used as synonyms by most theorists. So I will do the same in the following text.

25 In Berlin's (1969) words, "(b)y being free in this sense I mean not being interfered with by others."

Most liberals tend to endorse a negative conception of liberty, as their focus is on non-interference (Kelly 2005: 61, Carter 2012). Some even hold the view that liberalism is "a doctrine [...] of the scope of government power", with the result that "[a]ny genuinely liberal theory must offer an account of limits to public power" (Galston 2007: 290). This, again, does not mean that liberalism is only concerned with setting limits to state interference (see next section). However, this should be seen as a crucial ingredient of liberalism, and it will be the main challenge for the discussion on a potential policy of sufficiency that follows in the later chapters.

Finally, it should be noted that a related and also crucial notion within liberal theory is that of "rights"[26]. Liberties are translated into rights within liberalism, but reversely not all rights are liberty rights. According to a famous classification, the noun "right" is used to denote four different concepts: Claims, privileges (or liberties), powers, and immunities[27]. Still, most rights can be seen as expressing some active (for privileges and powers) or passive (for claims and immunities) freedom which their holders are entitled to (Wenar 2011). The liberties and rights of individuals are limited by the liberties and rights of others. Under which circumstances this is the case will be discussed later[28].

2.1.2 Equality

As opposed to the certainty apparent in the quotation by Gaus and Courtland above, some writers even disagree that liberty is the central value in liberalism (Gray 2004, Kelly 2005: 7). Instead, they hold that equality should be seen as the most fundamental value (e.g. Dworkin 1978).

The notion of equality also requires further definition. Leaving aside the most formal sense of equality, that is, the sameness of two things or people in a certain respect, which is just descriptive and not a value, the following distinction is often made: there is, on the one hand, the basic

26 The nature and justification of rights is subject to comprehensive discussion within liberal theory (cf. e.g. Martin 1998: 326, Wenar 2011), which I do not reflect here.

27 According to an analysis by Wesley Newcomb Hohfeld from the early 20th century (cf. Martin 1998).

28 See in particular the discussion on the "harm principle" in section 4.2, which states that within liberalism, the only legitimate reason for exercising power over individuals against their will is to prevent harm to others.

principle of equality of people's moral status, and on the other, the egalitarian question of to what extent certain goods or opportunities should be distributed equally in order to protect that basic principle (Dworkin 1978: 125, Kelly 2005: 9).

The former today is widely uncontested, not only within liberalism but also as a "minimum standard throughout mainstream Western culture" (Gosepath 2011). It demands that individuals be regarded as equal moral subjects and treated as equals, and not treated unequally due to, for instance, their sex or origins. As regards the second point, so-called egalitarian liberals argue in favour of a greater degree of equality in the distribution of items of any "currency" of justice (income, resources, opportunities etc.) (cf. Arneson 2013). This egalitarian principle competes with other distribution principles, such as need or the maximisation of overall utility (cf. section 2.3.3). This second type of equality is much more contested than the fundamental claim to treat individuals as equals.

It is easy to imagine cases where equality and liberty are in conflict. There is no agreement about a possible hierarchy of these two values central to liberalism. Even though liberty certainly is a core concept, many liberals consider the idea of equal dignity and respect as equally or even more important (Kelly 2005: 61). Whether and to what extent equality in distributions should sometimes trump a higher degree of freedom is highly contested. Many contemporary theories of liberalism combine equal basic liberties with the right to equal access to basic economic resources (Kelly 2005: 71). In the sense of a minimum standard, the principle of equality of individuals' moral status is adopted here, while the further-reaching idea of equality in distributions will be discussed in detail later in section 2.3.3.

2.1.3 Liberalism and neutrality towards the good

Another debate related to liberalism, which deserves some space as it is of particular importance for the subject of this study (cf. section 2.1.5), deals with the principle of state neutrality towards individual conceptions of "the good". In most characterisations of liberalism, the neutrality principle is taken to be an essential component. It states that in a pluralistic society, the state should not favour any particular conception of the good (life)[29] through policies or other means but should rather provide a neutral framework within which individuals can pursue their own potentially conflicting views of the good (Kymlicka 1989: 883, Wall/Klosko 2003: 6). Even though the mainstream of liberal theorists adopts some form of the neutrality principle, there are also voices that stated as early as in the 1990 s that "[…] it is hard not to believe that the period of neutralist liberalism is now over."[30]

Critics state that this allegedly crucial notion of liberalism has emerged only relatively recently and is thus not as essential to liberalism as many take it to be today (cf. Stephens 2001 a). Indeed, the explicit discussion of the issue dates back only to the 1970 s (Waldron 1989: 62, Goodin/Reeve 1989: 1, Wall/Klosko 2003: 2 ff.). While some state that long before this, liberals had adopted notions very similar to neutrality[31], others argue that in the 19th century some liberals held clearly perfectionist ideals (Wissenburg 2006 a: 142). Perfectionism is the antagonist to neutrality. Theories of perfectionism hold that there is an objective account of the good immanent in human nature and that policies should be designed accordingly to advance this good (Wall 2012). After all, neutrality may be a traditional notion—veiled in other concepts—and there may be counterexamples as well, since theorists have developed different variants of liberalism.

Discussions on the neutrality principle have in particular dealt with its justification, its scope, and with what exactly it is that is destined to be neutral. In terms of justification of the principle, it has been argued that the reasons for promoting state neutrality must themselves be neutral and not be based on values such as individual autonomy (Larmore 2003). Oth-

29 Apart from controversial ideals of the good life, there is no common understanding of what exactly should be included in the notion "the good" (Wall/Klosko 2003: 25).

30 Thomas Hurka in a review of "Beyond Neutrality: Perfectionism and Politics" by George Sher, Ethics 109 (1), 1998, cited after Arneson (2003: 192).

31 See, e.g., Goodin and Reeve (1989: 1): "[…] liberals have long been keenly aware of something very closely akin to the virtue of neutrality on the part of the police, the judiciary and the military: unless they dispense justice without fear or favour, liberal theory in practice simply does not work. Thus, while it is only in the last decade or so that the term has come into particular focus as a general issue, apparently similar notions have long figured largely in recognizably liberal political discourse."

ers insist that neutrality itself is a value (Waldron 1989: 72, Stephens 2001 a: 4 f.). At least, to some extent it already follows from the liberal norm of equal treatment that the state should not privilege the views of some citizens at the expense of others (Dworkin 1978: 127, Wall/Klosko 2003: 2). This plausibly justifies neutrality towards individual conceptions of the good, at least to a certain degree.

As regards the second point, the principle's scope, it is controversial whether neutrality should be applied only to the constitutional essentials or also extend beyond this, covering all policies (cf. e.g. Raz 1986: 112 f., Arneson 2003: 209 ff., Wall/Klosko 2003: 6). Much criticism has been brought forward on the idea of covering constitutional essentials only, as the term is neither clearly defined, nor is it obvious that "lower-level" policies, if non-neutral, might not have very substantial effects on citizens' lives as well, for instance in religious questions (Arneson 2003: 209). This position is therefore somewhat incoherent. A more plausible limitation of the principle's scope would be based on an evaluation of the degree of coercion[32] involved in a certain act of exercising political power (cf. Wall/Klosko 2003: 7), as coercion is, from a liberal point of view, unlikely to be accepted as an appropriate means to promote the good.

Finally, as far as what exactly it is that should be neutral is concerned, two positions are often distinguished: neutrality of the state's justifications for the policies it pursues versus neutrality of the consequences or effects of these policies (Kymlicka 1989, Waldron 1989: 66 f., Wall/Klosko 2003: 7 ff.; cf. also Arneson 2003: 193 f.). The idea of non-neutral effects of policies seems, at first sight, unrealistic, as virtually any policy has some adverse consequences for some people in practising their view of the good, and, in addition, there may be unintended or unanticipated effects. Taking this into account, a more realistic version of the neutrality of effects principle has been proposed, suggesting that those disadvantaged by policies could be compensated in some way. Still, most writers have endorsed neutrality of justification as the right formulation of the principle (Wall/Klosko 2003: 7 ff.).

As Wall and Klosko (2003: 21) note, in the past opponents of the neutrality principle have often focused on rejecting too simple versions of it, while critics of perfectionism have tended to advise against perfectionist visions which make it look hostile to tolerance and freedom. They there-

32 On the notion of coercion, see section 4.3.5.

fore recommend that new research on the neutrality-perfectionism debate should be more differentiated and address, among other things, the task of fathoming out the frontiers of compatibility: how much perfectionism can liberalism accept, or the other way round, what is the minimum of neutrality which is necessary in a liberal state? And: what kind of policies would be precluded by a plausibly formulated neutrality principle?

In line with the mainstream view of liberal theorising, I adopt the neutrality principle here as an element of liberalism, and the analysis which follows in sections 4.2 and 4.3 will be based on a neutral justification. This does, however, not necessarily rule out policies based on non-neutral grounds under all circumstances. More attention will be paid to this subject in section 4.4.

2.1.4 Categories of liberalism

Based on the various possible views on the constitutive elements of liberalism I have outlined, different types of liberalism are often distinguished. An important distinction relates to the two categories economic and political liberalism (see e.g. Hardin 1993)[33]. Both take liberty to be at least one of the core values; however, they focus on the protection of quite distinct spheres of liberty, and allow for different sorts of justification for limiting it.

The justification for state interference in individual liberty that will be used in the argumentation following later (cf. section 4.2) is in line with both variants, even though it targets political liberalism specifically. That said, I think it is still worth describing briefly the differences and linkages between the two. Many environmentalists outside the philosophical debate are very critical about liberalism (see also next section), but I am convinced that the bulk of their criticisms targets the economic variant, including some only allegedly essential elements of it.

The principal aim of political liberalism is to promote tolerance, personal freedoms, individuality, and the equality of people's moral status, es-

---

33 Other labels for distinct types of liberalism include: classical vs. modern or social (Wissenburg 1998: 73, 2006 b: 23), "mainstream American" vs. social (de-Shalit 2000: 83), utilitarian or welfarist vs. deontological (Sagoff 1988: 152, Hardin 1993: 122), or simply "old" vs. "new" liberalism (Galston 2007, Gaus/Courtland 2010). Interestingly, there are always two basic categories.

pecially with a view to tradition-related or religious restraints imposed by social groups or the state. This emanates from the assumption that various reasonable comprehensive doctrines, which are mutually exclusive but do not reject the basic characteristics of a democratic order, exist and can, in principle, exist in a society together (Rawls 1998). The debate on the neutrality principle thus relates in particular to political liberalism.

The programme of economic liberalism aims particularly at safeguarding property rights and free trade instead of state monopoly and protectionism, and, in its classical form, sometimes even conceives of liberty and private property as the same thing (Gaus/Courtland 2010). It is also argued that these economic freedoms are necessary to safeguard the more political freedoms (cf. Dryzek/Dunleavy 2009: 104). As opposed to what environmentalists would suspect, infinite economic growth (cf. section 2.3.5) is not a necessary ingredient of economic liberalism (Wissenburg 2006 b: 21, cf. also Dworkin 1978). Indeed, it tends to promote welfarist values (Hardin 1993: 135), and some classical theorists such as Adam Smith aimed at an increase in material prosperity with their ideas, but in Smith's days the societal potential of further economic growth was certainly different from the situation in developed countries today. Also, so-called liberal politicians today usually have economic growth at the top of their agendas (de Geus 2001: 32 f.), but "liberal" practice is not necessarily in line with liberal theory.

If, for the sake of simplicity, two sets of freedoms are distinguished between—personal (e.g. sexual orientation) and economic freedoms (e.g. freedom concerning one's property)[34]—it can be said that while liberals concur about the general endorsement of personal freedoms, they are more divided about the degree to which economic freedoms should be promoted or curtailed. Modern liberal theories, such as that developed by John Rawls, are concerned with combining the personal freedoms that protect the individual with the idea of social justice (Kelly 2005: 33), therefore allowing for more interference in the economic sphere than their classical counterparts.

34 This is suggested, for example, in the "Nolan Chart"—a diagram supposed to represent the political spectrum by allocating different political doctrines within a field spanned by two axes indicating the degree of economic and of personal freedom.

2.1.5 Liberalism and Green Political Theory

Traditionally, environmentalists[35] have been rather sceptical about liberalism, which was long seen as a reason for environmental problems and not as a theory strand that could deliver meaningful solutions. Some even seemed to be inclined to the idea that illiberal measures are the only way to deal with "the environmental crisis" (cf. Doherty/de Geus 1996, Meyer 2011: 360 f.). However, as stated above, this in part has to do with views of liberalism that focus on economic rights and freedoms rather than political ones, as well as with an alleged inherent link between liberalism and economic growth. On the other hand, there are also elements of liberalism that might conflict with environmentalists' demands and that can be factored out less easily. Among these elements, there are to mention in particular liberalism's focus on human beings rather than, for example, all animals or living organisms (cf. section 2.3.2), its tendency to be neutral between different conceptions of the good (cf. sections 2.1.3, 4.4.1, 4.4.3), as well as its central concern with protecting individuals from unjustified state interference (cf. section 2.1.1).

It has been only since around 1990 that a number of scholars have started to ask whether liberalism, and which of its variants, is compatible with sustainability or with stricter environmental protection (e.g. Achterberg 1993, Barry 2001, Baxter 2006, Bell 2002, 2006, Calder/McKinnon 2011, de Geus 2001, 2004, de-Shalit 1995, 2000, Hailwood 2004, Sagoff 1988: 146 ff., Stephens 2001 a, 2001 b, Vincent 1998, Wissenburg 1998, 2001 a, 2001 b, 2006 a, 2006 b). Following the bulk of literature by environmental theorists on the possibility of greening liberalism, this preoccupation has also started to incur criticism (cf. Meyer 2011).

Nevertheless, I think it is still of utmost importance for environmentalists to discuss the compatibility of environmental aims and liberal-democratic values, as for instance green parties are sometimes badmouthed as "prohibitors' parties" or put into the totalitarian corner by their opponents right away (cf. chapter 1). With regard to the subject of this study in par-

35 "Environmentalists" as well as "environmentalism" are used here in a broad sense to capture views that argue for meaningful environmental protection in general. Where the terms appear in this text, they therefore do not follow the distinction between "environmentalism" as rather shallow and "ecologism" as a deep and more genuine environmentalist standpoint, as suggested by Dobson (e.g. 2007) and followed by many others.

ticular, any attempt to restrict consumption for environmental reasons might be seen as a violation of the neutrality principle, and those in favour of it may be accused of aiming to impose their view of the good on others (cf. Coglianese 1998). Such accusations need to be taken seriously; this and the other aforementioned aspects of liberalism that might conflict with environmentalists' demands will thus be dealt with in later sections of this study (cf. sections 2.3.2, 4.2, 4.4.1).

2.2 Democracy

The meaning of "democracy" is less controversial than that of "liberalism".[36] It refers to "a method of group decision making characterized by a kind of equality among the participants at an essential stage of the collective decision making" (Christiano 2008). This equality implies that decisions will be made based on some sort of majority rule. "Collective decision making" means that the decisions under concern target the group and are binding to all of its members (Christiano 2003: 4).

2.2.1 Democracy and Liberalism

The requirement of equality among the group members points to an inherent link between democracy and liberalism (cf. Hayek 1960: 103). While this is probably the strongest, other links have been suggested in addition, such as liberty and autonomy in decisions over one's affairs, which are deeply affected by the sort of collective decisions taken in a democracy. It is argued that only when each individual has an equal vote in collective decision-making will there be a chance for an individual to be a master of their own life (Eckersley 1996: 222, Christiano 2008). However, there is also some tension between democracy and liberalism if the former is seen primarily as a procedure for collective decision-making and the latter as a programme dedicated particularly to limiting the scope of such decisions

36 Of course, there are also many different traditions and lines of thought within democratic theorising, some of which will be outlined below. However, the general meaning is not controversial in a similar way to that of liberalism, which would be a reason to say that "there is no such thing as democracy" (cf. the view of "the sceptics" on liberalism in Dworkin 1978).

in order to avoid a "tyranny of the majority". Indeed, liberal theorists were long sceptical about or even hostile towards democracy, fearing that it would lead to mob rule and denial of liberal individual rights (Dryzek/ Dunleavy 2009: 23). From a more community-oriented perspective, a common criticism about liberalism in this sense relates to its focus on the individual, which might be inadequate in ensuring the stable kind of communities needed in democracies (cf. Talisse 2005: 7 f.). Nonetheless, instead of characterising this as a tension, it can also be seen as complementary framework conditions, as has been expressed concisely by Talisse (2005: 82):

> *"[...] the political will, which emerges from the democratic/aggregative process, is constrained by the liberal principles. In this way, the individual is, on the one hand, protected by democracy from the oppression of tyrants and, on the other hand, protected from democracy by the rights and entitlements bestowed upon the individual by liberalism."*[37]

History shows that this second view of mutual complementation is more appropriate. Today, liberal democracy has become the most important state form (Dryzek/Dunleavy 2009: 18)[38]. The combination has become so common that "democracy" is even used as the short form of "liberal democracy" (Talisse 2005: 6, Freeden 2008: 22). Most contemporary theorists would say that democracy and liberalism are essentially connected with a view to basic values as well as, from an instrumental perspective, to ensure the best outcomes. Theorists such as Amartya Sen (implicitly) even conceive of both as a single entity, rejecting a primarily procedural view on democracy. Instead, Sen sees it as a complex and demanding system that requires the protection of liberties and legal entitlements in order to allow for adequate opportunities to present different positions, for instance in advance of elections (Sen 1999: 9 f.).[39]

---

37 Cf. also Hayek's characterisation of the relationship between liberalism and democracy (1960: 103).

38 Cf. also Freedom House's most recent report on "Freedom in the World", which classifies 45,6% of the world's 195 countries as "free" and therefore as liberal democracies. The share of so-called "electoral democracies" that do not all feature a sufficiently high level of civil liberties to be classified as "free" amounts to 64,1% (Freedom House 2015 a, 2015 b).

39 Cf. also the summary of positions on the relationship between democracy and justice by Dowding et al. (2004: 5 f.) and the minority position of Arneson (2004) in the same volume.

2.2.2 Empirical characteristics and theories of the democratic state

On an empirical view, modern (liberal) democracies are characterised by a number of common features, in particular: the sovereignity of the people; periodic, free elections based on fair competition among those running for an office; a certain degree of political participation and political and social pluralism; limits to the exercising of power and authority through the separation of powers; respect for the rule of law and fundamental civil liberties; and a certain degree of social welfare (cf. Frevel 2004: 57 ff., Dryzek/Dunleavy 2009: 18).

However, modern democratic states also vary considerably in their institutional settings and democratic culture. Some have more representative elements, while others entail more procedures of direct democracy, some rely more on majority decision-making, while others strive more for consensus, and they vary in their electoral systems and degree of centralisation. This is partly due to their historical development and cultural and social characteristics, but partly also to different theoretical approaches to the idea of democracy with their respective images of human beings and the state, which have influenced democratic practice (Frevel 2004: 67).

There are various competing theoretical approaches to democracy, and if one compares different readers on the subject, it is not even clear what should count as a genuine democracy theory and what are rather theory debates which are interpreted in the light of their influence on democracy. Such theoretical approaches focus on different aspects and can either be explanatory in the sense that they try to shed light on how things actually work in a democracy, or normative in that they suggest how things ought to work, or both.

Two of the most influential democracy theories are pluralism and elite theory. Pluralism emphasises the diversity of society and stresses the beneficial effects of such a diversity. Based on this, pluralists accept the unavoidability of conflicts of interest and stress the importance of compromises. Pluralists hold that citizens with common interests organise in groups, which have a major role in representing the diverse interests in political discussions, and that indeed all political decisions can be traced back to group interests (Waschkuhn 1998: 19 f., 76 ff., Frevel 2004: 71 ff., Dryzek/Dunleavy 2009: 35 ff.). Pluralism is historically rooted in liberalism, as its focus on competing interest groups corresponds to liberalism's focus on free individuals under the assumption that in complex modern societies single persons do not compete individually with each other to achieve the best democratic results, but that they are organised in interest groups (Waschkuhn 1998: 83 f.)[40].

Elite theory builds on empirical studies which revealed that most citizens show little interest in politics and have rather poor knowledge of political affairs. It thus assumes that not all citizens form groups, but that democracies are run by elites and—according to some theorists—also need this kind of leadership in view of the limited knowledge of citizens. Citizens' contribution is limited to the process of elections that serve to select and legitimise the governing elite (Waschkuhn 1998: 20 f., Frevel 2004: 70, Christiano 2008, Dryzek/Dunleavy 2009: 57 ff.).

Both the pluralist and the elitist approaches are criticised by those in favour of a more participatory democracy, who express their disapproval of reducing political interaction to just politicians and association representatives (Waschkuhn 1998: 26). They argue that the existing political apathy should be interpreted as a failure of the democratic institutions rather than as an unavoidable effect (Dowding et al. 2004: 2). However, broad participation of citizens in all issues is seen as unrealistic, given the costs and time this would require, as well as the cognitive capacities needed to deal with complex political affairs and the limited time average citizens have available for this, as political activity always competes with other individual interests. It thus seems necessary to determine more specifically the areas where participation can be successful (Waschkuhn 1998: 27, 41 ff.).

2.2.3 Deliberative democracy

As a successor to participatory theories[41], the perspective of democracy as being "deliberative" (or "discursive") has had a steep, upward career since the 1990 s, even though the idea itself is much older (Elster 1998: 1, Gut-

40 Cf. Fraenkel (1973: 202) for a different interpretation which highlights the tension between liberalism and pluralism that arises due to their different focus on individuals versus groups. Cf. also Dryzek and Dunleavy (2009: 35 ff.) on the interlinkages of pluralism and liberalism.

41 See Dryzek and Dunleavy (2009: 213). Cf. also Pellizzoni (2010: 171), who contrasts deliberation and participation with a focus on those who make a decision: "deliberators cannot become actual decision-takers without undermining the role of democratic institutions—the deliberative process is mostly consultative rather than participatory." In practice, many instruments used to increase public partici-

mann/Thompson 2004: 8 f.). Some even speak of a "deliberative turn" in political theory around 1990 (Talisse 2005: 85, Dryzek/Dunleavy 2009: 215). I will go somewhat deeper into this perspective on democracy than the other ones mentioned so far, as it seems to be of particular relevance for the subject of this study, and as it was also taken up by several environmental theorists (e.g. Dryzek 1992, J. Barry 1996: 118, Hayward 1998: 97 ff., de-Shalit 2000: 141 ff., 163 ff., Pellizzoni 2010).

Dryzek and Dunleavy define deliberation as "a particular kind of communicative process in which individuals reflect upon their own views in the light of what others have to say, ideally in contexts free from coercion, manipulation and deception" (2009: 215). More precisely, within this process of reflection individuals need to give each other "reasons that are mutually acceptable and generally accessible"[42] instead of merely pointing to the alleged will of the majority or generic explanations, such as the explanation that something promotes the national interest (Gutmann/Thompson 2004: 3, 7). It is thus the concept of justification which provides the core of deliberative democracy (Cohen 2003: 21, cf. also Christiano 2008).

The idea of democracy being a communicative process, relying on an exchange of arguments and critique seems so natural that it inspired Nida-Rümelin to say that "in this sense democracy is always deliberative. [...] We live in a deliberative democracy or we do not live in a democracy" (2006: 41)[43]. However, democracies actually exist—such as the "electoral democracies" identified by Freedom House (2015 a, 2015 b)—that are not deliberative, because they are not liberal democracies that allow for free expression of opinion. Without such fundamental rights, which are provided for in a liberal society, free and open deliberation cannot occur. It must thus be assumed that Nida-Rümelin's statement refers to liberal democracies only.

The focus on arguments distinguishes deliberative democracy from other modes of communication available in a democracy, such as bargaining (Elster 1998: 5 f., Pellizzoni 2010: 169). Further, deliberative democracy is often contrasted with majority decision-making based on the aggrega-

---

pation are also rather consultative (e.g. hearings), so there appears to be some overlap.

42 Note that there are discussions about the possibility of such reasons, which are acceptable to all in a pluralist society, and the difficulties of distinguishing such acceptable reasons from others (e.g. Cohen 2003: 22 f., Bohman/Richardson 2009).

43 My translation.

tion of preferences in elections, opinion surveys, or direct democratic decisions (Cohen 2003, Gutmann/Thompson 2004: 13 ff., Pellizzoni 2010: 169, Dryzek/Dunleavy 2009: 216; cf. also Elster 1986). As Gutmann and Thompson (2004: 13) state, the deliberative conception of democracy focuses on the justifications that deliberators give each other for their expressed preferences, while the aggregative conception of democracy takes individual preferences as given and does not require any justification. This sharp theoretical contrast between the two conceptions, however, seems to disappear in practice. As indicated by Nida-Rümelin, indeed before any majority decision that affects the public is taken (except of course for secret, non-public ones), be it in parliament or by the citizens themselves, there is always an exchange of arguments. At least it is hard to think of a democratic vote where talking about the subject of the vote before is not allowed, leaving individual preferences undisturbed by others' views (cf. Elster 1998: 3)[44]. On the other hand, in the deliberative conception it is also clear that even a careful evaluation of competing justifications does not necessarily influence individual preferences and voting decisions, since other reasons that speak in favour of a certain position might be more important.

The place where deliberation is said to occur does not only include the formalised bodies for discussion in a democracy, such as parliaments, but, according to some theorists, also the public sphere in general, which can cover informal arenas such as the media and "conversation in cafes and pubs" (Dryzek/Dunleavy 2009: 219 ff., Pellizzoni 2010: 169 ff.). In addition, there can be special "deliberative arenas" for citizens to meet and discuss which have been facilitated by public authorities and would also fit into the category of instruments for public participation.

Public deliberation in a democracy is widely considered to be a good thing, as it increases the legitimacy and the quality of decisions if all parties concerned[45] can participate in the deliberative process and express their arguments (Dryzek/Dunleavy 2009: 217, Pellizzoni 2010: 170), par-

---

44 Except for the case of mere "electoral" democracies mentioned above. When there is no right to free speech, citizens or parliamentarians are of course limited in their ability to exchange arguments and deliberate openly over the current issues of concern.

45 Of course with some exceptions such as certain handicapped people, future generations, and animals, as e.g. green theorists would be quick to add. However, these can at least be represented by respective organisations that speak on their behalf.

ticularly as it asks for arguments that are accessible to others (Gutmann/Thompson 2004: 3 ff., see above). Further, as claimed by some, deliberation promotes the common good because of the impossibility of bringing forward selfish arguments in public debates with regard to public concerns, and because at the same time it is likely that, if only non-selfish, generally accessible arguments are accepted, one will finally come to endorse these arguments which one expresses (cf. Elster 1986: 113).

On the other hand, the deliberative process has also weaknesses: problems of scale, since not everyone affected by an issue can take part in discussing its solution; withdrawal of uninterested citizens and self-selection of those participating in debates; the vulnerability of the deliberation process to manipulation due to potential uncertainty about the real motivation of participants or selected experts (Stokes 1998, Pellizzoni 2010: 171). In addition, the deliberation process is limited by the time available for discussion with the result that in the end, an imperfect decision may have to be made based on "a little discussion" (Elster 1986: 115). There are solutions to some of these weaknesses that can at least mitigate the respective consequences. For instance, with regard to problems of scale, the election of representatives or their selection by lot are common practices (Dryzek/Dunleavy 2009: 217 f.).

The "deliberative turn" has also influenced liberal theory (Talisse 2005: 85 ff.). The question that is of interest here is: is the view of democracy as being deliberative compatible with liberalism? Some theorists suggest that there are problems of compatibility. Talisse, for instance, argues that while liberals have in principle embraced deliberation and supplemented their views with this strand of theory, they still place constraints upon the issues that are open to deliberation: namely, they do not accept their basic liberal principles being called into question, which Talisse sees as an inconsistency in liberal theorising about deliberative democracy because these constraints are not generated by the deliberative process itself (2005: 85 ff.). However, going back to the general relationship between democracy and liberalism as outlined above, this seems to point again to the "tension", or better, the complementary framework formed by these two (which is not characterised as problematic by Talisse, see above), rather than to any specific problem of *deliberative* democracy and liberalism[46]. It thus seems that there is no fundamental contradiction between deliberative democracy and liberalism, but rather that they focus on different aspects of collective decision-making in a liberal democracy, and even though they are not free of tensions, complement each other with regard to these aspects.

Summing up, for the purposes of this study I understand democracy (within a liberal-democratic system) as a method of collective decision-making that is based on equal voting rights and characterised by deliberation among the participants, which consists in an exchange of justifications they give each other to support their respective positions.

2.2.4 Democracy and Green Political Theory

With regard to green political theory, something similar can be said for democracy as it can for liberalism, even though not to the same extent: in the past some theorists have argued that drastic, anti-democratic measures are needed in order to save the planet, as the majority is supposedly too lethargic or disinterested to deal effectively with environmental issues. Some environmental activists, even if they constitute a rather marginal group, have chosen to use violence as their means as allegedly debate and majority vote do not lead to the desired results (Radkau 2011: 409 ff.).

Indeed, there seem to be a number of problems with democracy that make it difficult to adopt serious environmental policies. One is that short-term individual (human) preferences dominate the agenda, while environmental problems often unfold their effects only in the long term (Eckersley 1996: 215, Wissenburg 2006 b: 21). Future generations cannot participate in democratic decision-making processes to express their interests, and thus only limited attention is given by decision-makers to their fate (Ball 2006: 131, de Geus 2001: 21). This is in a similar way the case for those who are affected by pollution emanating from a country but who

46 See also Gutmann and Thompson (2004: 9, 26) on an alleged contrast between Rawlsian and Habermasian theory in this sense: "Habermas is said to favor democratic deliberation over individual rights, and Rawls, rights over deliberation. But on the more careful interpretations of their theories, neither Habermas nor Rawls defends a purely procedural or purely substantive conception of democracy. […] The convergence between Habermas and Rawls suggests that the most compelling theories of deliberative democracy combine both substantive and procedural principles." Cf. also Habermas (1996).

live outside its national borders, with the result that there is a tendency to externalise environmental costs (Eckersley 1996: 214). In addition, compromise is an essential element of democratic procedures, so that the green programme, if adopted at all, might end up "in dribs and drabs and pieces" (Ball 2006: 133).

Thus, similar to the mentioned tension between liberalism and democracy, there appears to be a tension between democracy and the environmentalists' programme. While democracy requires a specific procedure, environmentalists advocate certain outcomes, and the procedure alone cannot guarantee the desired outcomes[47], as Goodin has argued (1992: 168; cf. also Saward 1993). This statement has triggered a discussion about the relationship of environmentalism and democracy, in particular on the question of whether there might still be anything inherent in democracy that makes it more likely to lead to environmentally sound outcomes than other systems, or, respectively, whether environmentalists' demands as such make them democrats (see the contributions in Doherty/de Geus 1996; de-Shalit 1997, 2000: 130 ff., Ball 2006).

Several writers hold the view that environmentalism is still more in line with democracy than with authoritarian systems (J. Barry 1996: 116, Ball 2006) or, put reversely, that democracy is more likely to protect the environment than any other system (Hayward 1994: 209, de-Shalit 1997). In this sense, one could argue that some environmental protection is needed to safeguard the preconditions of democracy. However, this would also be the case for most other modes of decision-making, as Dobson has pointed out (1996 a: 134 ff.). Dobson sees it, instead, as more convincing to link green thinking and democracy through their "common concern for autonomy" (1996 a: 146). Some have even tried to build a "green democratic right" to a clean environment into the principles of democracy (cf. Saward 1996, Eckersley 1996). However, such a right would essentially be a liberal right emanating from the principle of avoiding harm to others[48] and not

---

47 The relationship between democracy and justice is also sometimes described in this vein: it has been argued that "there is nothing inherent in democracy that necessarily makes it just. Democracy is a procedure for formally capturing the views of the citizens and translating them into outcomes. That procedure has only tangential connections to the outcomes being just." (cf. Dowding et al. 2004: 5, describing the position of Brian Barry).

48 See section 4.2.1.

a "democratic right"[49]. With again a different focus, it has been argued that, for strategic reasons, a structural and lasting solution to environmental problems can only be based on democratic decision-making, since the related changes in lifestyles are so significant that they would never be permanent under authoritarian rule as they would not be supported by citizens (Achterberg 1993: 83).

Empirically, there indeed seems to be a positive relationship between democratic government and the level of environmental protection (J. Barry 1996: 116). It can be observed in many countries that democracies are not necessarily unreflective of the long-term perspective if citizens have an interest in such a perspective and can freely express their opinion with regard to it (Birnbacher 1988: 258 ff.). Still, there certainly is much left to do in existing democracies from the environmentalists' perspective.

The reference to free expression of opinion points to the discussion culture prevalent in liberal-democratic societies. Several "green" theorists have therefore embraced deliberative democracy as their preferred view on democracy because they think it serves best in bringing about solutions to environmental problems[50]. It is argued, among other things, that deliberative democracy, as it requires people to provide justifications and arguments, encourages citizens to reflect on their preferences and potentially prompts them to adopt more environmentally sound ways of living. Achieving such a transformation based on conviction would be more effective and long-lasting than being told to adopt a different lifestyle (J. Barry 1996: 122). However, as Hayward (1998: 103) has pointed out, even though a discursive clarification of the interests of those participating in a discussion can be expected to lead to more "enlightened" interests, this does not necessarily entail substantive environmental commitments. The same concern is raised by Dryzek (1992: 38 ff.), who at the same time suggests that a number of characteristics of the deliberative (or discursive) process at least make it more likely to change people's preferences into more environmentally sound ones. In particular, he mentions the need to formulate generally accessible arguments that, in an environmental context, will often refer to public goods and thus help to prevent the "tragedy of the commons".

49 Cf. also the related criticism by Dobson (1996 a: 142).

50 For an overview of other perpectives on democracy that "green" theorists have adopted, see de-Shalit (2000: 135 ff.).

Another reason why environmentalists should adhere to deliberative democracy relates to the flow of information related to environmental decision-making. Only in this way, as de-Shalit (2000: 163 ff.)[51] emphasises, can governments be pressed to disclose detailed information on hazardous waste, plans for development, or potential ecological disasters. Because of the deliberation, the flow of information also moves in the other direction —from citizens to the centre, where policy decisions are taken. At the same time, the knowledge of the public on environmental issues is improved, which might otherwise be very limited (de-Shalit 2000: 165).

A related point deals with the assessment of risks and uncertainty (cf. Box 2 in section 2.3.4), which play an important role in the evaluation of environmental problems. Deliberative democracy might be simply the best way to find solutions under conditions of risk and uncertainty, as knowledge production and expertise alone might not be sufficient for this task. As Pellizzoni (2010: 166) states, because of the complexity of many environmental problems

> *"[f]acts become soft and values hard; cognitive uncertainties blur with ethical uncertainties. [...] The higher the stakes and the deeper the political controversies, the deeper the scientific uncertainties or the higher the levels of certainty required to decide."*

Uncertainties therefore become a political matter, which requires the involvement of those affected by a decision (de-Shalit 2000: 165 ff., Pellizzoni 2010: 168, cf. also J. Barry 1996: 118 ff.).

At the same time, there are also limitations to deliberation as a solution to the challenges encountered in the environmental field (Pellizzoni 2010). Whether (deliberative) democracy will lead to the outcomes favoured by environmentalists in the long term remains to be seen and depends on whether they will succeed in convincing a sufficiently great number of people of the necessity of these outcomes (cf. also Bell 2002: 721, Achterberg 1993: 83). There are a number of arguments in favour of deliberative democracy's success in this task, but there is no certainty about the empirical result in the end. I will get back to this question in a later chapter (cf. section 4.4.4).

---

51 Actually, de-Shalit argues in favour of "deliberative participatory democracy"; yet he mainly seems to understand participation (besides voting) as taking part in deliberation processes (cf. de-Shalit 2000: 144).

Now that the basic tenets of liberal democracy, as understood for the purposes of this study, have been outlined, the next section will present what the "green" outcomes argued for in this study comprise. More explicitly, the next section will provide a thorough analysis of the concept of sustainability. I will deal with this concept and its related controversies at somewhat greater length. One reason for this is that I postponed dealing with some topics that are also closely related to liberalism and democracy, such as the issue of justice, to the chapter on sustainability, as this represents the more specific perspective here. Another reason is that sustainability discourse forms the basis of the main concern of this study, the sufficiency debate.

2.3 Sustainability

2.3.1 Introduction

The roots of the term "sustainability" lie, as is usually stated, in German forestry in the 18th century. "Sustained"[52] referred to a way of managing forests that allowed for logging only as much wood as could grow within the same time through new plantations (Grober 2002, Kopfmüller et al. 2001, Ott/Döring 2004). A comprehensive historical overview of the origins and development of the term itself and of the sustainability discourse is omitted here, as this is available elsewhere (see especially Grober 2002, 2010; see also Kopfmüller et al. 2001, Ott/Döring 2004, Michelsen/Adomßent 2014). I will only very briefly summarise the more recent developments within the sustainability debate before entering the conceptual matter.

The 1980 s, 1990 s and early 2000 s brought a number of milestones for the sustainability debate, in particular the so-called Brundlandt report (WCED 1987), the *Agenda 21* and Rio Declaration (UNCED 1992 a, 1992 b), and the documents from the Johannesburg Summit ten years later (UNWSSD 2002 a, 2002 b), as well as numerous national and supranational level reports, policy declarations, and strategic papers[53]. By the turn of the millennium, the term sustainability, or sustainable development, had

52 In German: *"nachhaltend"*.

53 For example, there are sustainability strategies on a European Union level (EC 2001 and later reviews) and in Germany (Bundesregierung 2002).

become very popular but very unspecific, too (Grunwald 2004). This lack of concretion was and still is perceived as problematic by many sustainability theorists. Environmental economist Herman E. Daly stated that even if nobody is sure of what the term means, everyone likes it: "At least it sounds better than 'unsustainable nondevelopment'" (Daly 1996: 1). Brand and Jochum noted in 2000 that nearly everything could be labelled as "sustainable", with the result that the concept had become a "list of wishes" onto which each interest group could place its non-negotiable interests (2000: 75). Or, more negatively, it has been claimed that sustainability has become "one of the most polluted concepts" in the recent past (de Geus 2001: 20).

While among the general public the term is still perceived as a blurry concept, the debate has moved forward since then in academic circles. In the USA and internationally, the emergence of the new research field "sustainability science" has been announced (Kates et al. 2001, Clark et al. 2004, Clark 2007, Levin/Clark 2010, Heinrichs et al. 2016). However, this research field comprises a multidisciplinary research agenda for sustainability rather than helping to clarify the concept itself. Numerous scientists have suggested definitions of sustainability, but there is no commonly accepted one (Parris/Kates 2003, Kates et al. 2005)[54]. What is clear is merely that, first, sustainability deals with keeping something—usually with a focus on environmental conditions—(sufficiently) intact and, second, that this should be done because of justice considerations in particular, but not only, towards future generations. Approaches differ in the emphasis they put on either of the two aspects[55]. In view of the countless definitions that have been proposed and the numerous sub-debates held under the umbrella of sustainability, it has been argued that a definition of such a complex concept should be avoided and instead, the development of a sound theoretical framework should have priority (Ott 2004: 39, Christen/Schmidt 2011, see also Daly 1996: 2). In the following, I will outline some theoretical approaches.

---

54 The Brundtland Commission's definition (cf. section 2.3.3) is indeed quoted frequently (and is also quoted below in this chapter), but it has also been criticised as being much too vague.

55 Cf. the map of sustainability concepts proposed by Hopwood et al. (2005), which spans the axes of environmental concerns (none/techno-centred/eco-centred) and socio-economic well-being and equality concerns (inequality/equality).

Dimensions and "Three Pillars Model"

An early and rather simple model is the "Three Pillars Model" or similar representations that embrace several "dimensions" that are to be considered jointly and the demands of which are to be balanced in order to achieve sustainability. Usually the three dimensions environment, economy, and society are used, which became popular after the Earth Summit in Rio in 1992 (Grober 2010: 21). In the Three Pillars Model each dimension represents a pillar of a temple, which would collapse if any of the pillars were missing or were smaller than the other ones[56]. The concept is not explicitly contained in the Brundtland report, but it was used later because it allegedly is the best way to reflect the aims contained in the report (SRU 2002: 68). Besides the three dimensions, proposals to extend the model with an additional political-institutional or cultural dimension have been made (see Kopfmüller et al. 2001, Hawkes 2001, Kuhn 2006, Michelsen/ Adomßent 2014: 30 f.). The model illustrates the idea that environmental concerns cannot be dealt with separately from social and economic ones and therefore a policy of integrating the three (or more) dimensions is necessary (Enquete-Kommission 1998: 18).

The model has been subject to two strands of criticism. One deals with the number and label of the dimensions. In this sense, it has been stated that the three dimensions lack theoretical foundation. It seems that the model's success is rather due to the beauty and intuitive logic of the number three, which only represents the good idea that stands behind it: that the world is complex, one-sidedness is always a problem, and actions supposed to provide a certain benefit somewhere need to be considered in terms of their potential impacts on other areas. The three thus seems to be an arbitrary invention, and in this sense the number of the dimensions is of limited importance from a theoretical viewpoint as long as the idea they convey is acknowledged.[57]

---

56 Other representations of the same idea display the three dimensions as corners of a triangle with a surface area to be maximised, overlapping circles, etc. In the international discussion, the idea has gained attention especially as the "Triple Bottom Line" in corporate reporting on sustainability performance (cf. Elkington 1997), where the dimensions are also known snappily as the "three P's" of "people, planet and profits". The bottom line is the result at the end of a profit and loss statement.

57 For more practical concepts that aim to operationalise sustainability in the form of measurable indicators, this division into different areas may, however, have an im-

The other strand of criticism, however, is more fundamental and deals with the weighting of the different dimensions. The three dimensions model implies that the environment, the economy, and social concerns should have equal weight, so that greater importance is not attached to one of them above the others. This makes it acceptable for almost everyone, especially if the mere postulate of considering the three dimensions equally is not followed by concrete action. The model is thus blamed for the blurry image of the term sustainability, because it allows for everything which is good to be loaded on the "sustainability bandwagon" so that it becomes a "landfill dump" for everyone's wishes (Goodland/Daly 1996: 1002; cf. also Brand/Jochum 2000: 75, SRU 2002: 68, Ott/Döring 2004: 35 f.).

As there is some tension between the three dimensions, the outcome is, in cases of conflicting interests, often a compromise or a result of prioritisation. According to the German Advisory Council on the Environment (SRU), the extension of the term sustainability, which originates from environmental discourse, to three dimensions often leads to prioritising short-term economic targets rather than environmental concerns. Economic aspects, however, are already incorporated well elsewhere into politics and society (SRU 2002: 160). The postulated equality of the three dimensions would, if implemented in practice, actually enhance the current status of environmental concerns. However, in fact this demand is often not met but instead, economic concerns get priority over, for instance, requirements of nature conservation in protected areas, such as national parks, and the model thus weakens the environmental realm (Ott/Döring 2004: 36). Therefore, detractors of the three dimensions model demand a return to strengthening the environmental dimension (SRU 2002: 68, 160) or shifting the focus of attention away from economic concerns (Luks 2007).

The focus of this study indeed lies on the environmental dimension and it tends to subsume the economic one under social concerns, as a "healthy" economy, whatever that means, is not seen as an independent goal but as something that is supposed to serve human ends. Environmental concerns are not prioritised over social ones, but the problem dealt with —excessive burdens on environmental carrying capacity—requires taking

---

pact on the result—for instance, if the target is to achieve a certain value in each of these areas.

a primarily environmental perspective, while the consequences, if the problem should further materialise, will inevitably also include severe social impacts (cf. chapter 1).

Other models and theories of sustainability

Beyond models using the three dimensions, other models as well as comprehensive theories proposed within sustainability discourse feature a much more diverse set of issues. An early attempt to organise these was presented by Dobson (1996 b). After surveying a large amount of literature on sustainability with regard to the questions being asked in these texts, he identified four conceptual questions that any theory of sustainability must answer: What is to be sustained? Why? Who or what is/are the object(s) of concern? To what degree is substitutability between human-made and natural capital allowed? Included in these are two implicit questions related to justice: What is to be distributed? And among whom? (Dobson 1996 b: 407). Dobson then sorted the different answers given to these questions into four groups, obtaining a typology of sustainability conceptions.

Similar to this, Christen and Schmidt (2011) proposed a "meta-concept" consisting in a framework with elements that, according to them, any substantive conception of sustainability must contain. First, these elements or "modules" include the "sustainability problem", that is, the dilemma of the necessity of improving the quality of life of those who are currently poor on the one hand, and the fact that this might conflict with the realisation of the same aim in the future. The framework further contains two modules representing principles of justice, on the one hand, and the principle of integration of the environmental with other domains, on the other. The next stages consist in a module representing criteria for sustainability and, finally, a module dealing with transforming sustainability into practice.

Approaches such as those developed by Dobson and Christen and Schmidt provide a formal framework only and do not give an answer to the questions related to the content of sustainability (Christen/Schmidt 2011). There are, however, also many substantial contributions to sustainability theory that do give such answers. A large part of these, in particular the earlier contributions, are based in economic theory (cf. Dobson 1996 b: 408) and intend to frame sustainability as a neutral paradigm to be evaluated in terms of (allegedly) value-free economic measures and up-to-date scientific knowledge. Further, two substantial and explicitly normative conceptions of sustainability were developed in the German language in the early 2000 s, namely the *Integrative Concept of Sustainable Development* (Kopfmüller et al. 2001, Kopfmüller 2006) and the *Theory of Strong Sustainability* (Ott/Döring 2004).[58] Both are too comprehensive to summarise here, but their ideas will be considered in the conceptual discussion of sustainability below.

An objective or normative concept?

As these different theoretical approaches to sustainability show, there is disagreement about whether the concept can be framed strictly in "objective" terms by reference to scientific findings about environmental carrying capacity[59] (Hueting/Reijnders 1998), or whether it necessarily also contains elements that are defined by reference to values (Kopfmüller et al. 2001: 126 ff., Parris/Kates 2003: 581 f., Clark et al. 2004, Beemsterboer/Kemp 2016). In principle, I agree with the latter that sustainability is not a purely objective concept. Normative elements are not only present in the debate on justice and the question of who the objects of concern for sustainability should be, but to some degree also in the determination of environmental carrying capacity, as well as with a view to possible target conflicts between different aims. This controversy will be reflected again, particularly in section 2.3.4 when it comes to the question of "what is to be sustained". However, I also think that it is possible to formulate a "relatively objective" *minimum* conception of sustainability, which is loaded with as few values as possible and is objective in the sense that it is necessary to safeguard human needs and to avoid harm to human beings, and does not take a position with regard to further-reaching requirements that would be more clearly based on values and involve greater controversy. As will be seen later on, sustainability as presented here will be framed as such an objective minimum conception, which is determined by thresholds of need and harm, while not denying the existence of further-reaching, normative aspects of sustainability.

58 For an English-language summary of both conceptions, see Schultz et al. (2008).
59 For the term "environmental carrying capacity", see section 2.3.4.

Relevant aspects of sustainability

Considering the different questions and aspects raised within the models and theories of sustainability mentioned above, I have identified a set of issues from the sustainability context that are of particular relevance for the purposes of this study. The analysis of sustainability in this chapter thus deals with the following debates: whether sustainability should take an anthropocentric stance or include more than human beings in the community of morally relevant subjects; what justice conception should be adopted, also with regard to global justice and between generations; the role of natural capital and the possibility of its substitution ("weak" vs. "strong" sustainability); and to what extent sustainability should or can be combined with some concept of development or economic growth.

2.3.2 Anthropocentrism and physiocentrism

A fundamental discussion in environmental ethics deals with the question of who or what should have intrinsic value as opposed to purely instrumental value (cf. de-Shalit 1995, Dobson/Lucardie 1993, Dobson 2007: 29, Hailwood 2004, Hayward 1998, Humphrey 2002, Kopfmüller et al. 2001: 152 ff., Ott/Döring 2004: 53 f., 165-168, Wissenburg 1998: 91-114).[60] Related to this, it is questionable who and what should be included in the community of morally relevant subjects or objects. Sustainability theory must deal with this "inclusion problem" because the answer to it determines whose entitlements are under discussion (Ott/Döring 2004: 53). Besides anthropocentrism, that is, the viewpoint that only human beings should matter, various positions that can be summarised under the header of "physiocentrism" have been argued for. These are, from least to most comprehensive: intellectualism (including all intelligent creatures), pathocentrism or sentientism (sentient beings), zoocentrism (animals), biocentrism (organisms), ecocentrism (nature or ecosystems), and

60 A sub-discussion to this has emerged on the question of whether giving nature or non-human beings intrinsic value is or is not compatible with liberalism (cf. e.g. Bell 2006, Baxter 2006, Hailwood 2004, Stephens 2001 a, Wissenburg 2001 a). This debate is not reflected here as there seems no need to go more into detail due to the approach taken (see further below).

holism (Kopfmüller et al. 2001: 156 ff., Ott/Döring 2004: 165, Wissenburg 1998: 107).

Besides the distinction between these positions, it also has to be specified what each of them would imply: if, for instance, sentient beings were included in the moral community, this does not necessarily entail they should have equal rights as humans, given the cognitive capabilities needed to exercise, for instance, political rights.[61] On the other hand, an anthropocentric perspective does not entail that *no* rights or entitlements are assigned to non-human beings. Therefore, also from an anthropocentric view non-human beings do not necessarily *only* have instrumental value (Hayward 1998: 47), and human beings do not necessarily have to be privileged generally in all cases of conflict of interest even when only "trivial" human needs are involved (cf. Humphrey 2002: x).

In the context of this study, an anthropocentric perspective which is "enlightened" in this sense (cf. Schultz et al. 2008: 477) is adopted, even if I have strong sympathies with theorists who put their efforts into arguments in favour of including more than human beings into the moral community. The reasons are more of a pragmatic sort rather than founded in theory: first, I am convinced that there still remains a lot to improve in terms of sustainability which will serve human interests *and* will be beneficial for other creatures at the same time. There are certainly plentiful conflicts of interest between human beings and nature when it comes to particular cases, for example preserving an area of land or making use of its resources. These are, however, often also conflicts of interest between different groups of human beings, as many have an interest in preserving nature reserves and species. A second reason is that an anthropocentric view is less controversial than any physiocentric view—probably not among environmentalists, let alone animal activists, but among the general public. It therefore allows for the broadest possible agreement with the arguments put forward in this study.

61 Nonetheless, some argue that certain animals could be entitled to something that comes very close to this, which is that they should, in principle, not only have fundamental but also citizen rights; as animals cannot, for instance, go to vote in elections, this could be done for them by ombudspeople or other representatives (cf. Donaldson/Kymlicka 2011: 101 ff.).

2.3.3 Intra and intergenerational distributive justice

The most famous definition of sustainable development was proposed in the report by the World Commission on Environment and Development ("Brundlandt Commission") in 1987: "Sustainable development is development that meets the needs of the present without compromising the ability of future generations to meet their own needs." (WCED 1987: I/2/1[62]). It claims that a particular desirable state, in which each generation is able to meet its needs, is to be sustained over time. Meeting needs requires, among other things, a basic stock of material and immaterial goods or resources. Hence, sustainability deals with the distribution[63] of goods, or something similar, and postulates that they should be distributed in a way that human beings, regardless of where they live on the globe and in which generation, are able to meet their needs.[64]

Distributive justice is already an extensively discussed subject with regard only to human beings who are currently alive. It does not only deal with the justice of different distribution procedures, but also of the resulting distribution patterns. Some theories put an emphasis on the distribu-

---

62 As the Brundlandt Report does not have page numbers, citations are provided here as follows: part (Roman numerals) / chapter / number.

63 Some argue that sustainability should also deal with other domains of justice. Christen and Schmidt, for instance, state that it should incorporate a conception of "political justice", "because certain goods that cannot be distributed must be ensured, e.g. the right to participate in legal systems" (Christen/Schmidt 2011: 5). While it should be noted that it is controversial whether rights and liberties can be distributed or not (cf. Dworkin 1987, Gosepath 2011), such elements of formal or procedural justice are, in my view, not at the core of the sustainability idea. As stated above, sustainability deals with the ability to sustain a certain desirable state into the future. Political rights do, in principle, not require material resources, so the possibility of granting them to future generations (or to everyone globally at a given moment) does not depend on their realisation today (or in a certain place on the globe). Of course, in practice developments in history with regard to political rights have an effect on the future, but they cannot generally preclude the realisation of a certain set of such rights later (as opposed to a certain pattern of distribution of material goods). In order to avoid everything that is considered a good thing being "loaded onto the sustainability bandwagon" (Goodland/Daly 1996: 1002), I therefore suggest keeping political rights and similar things out of the concept. Sustainability certainly becomes more convincing if it is *combined with* a conception of political rights and liberties (cf. sections 2.1, 0), but these do not have to be elements of sustainability itself.

64 On the term "needs", see Box 1 below.

tion procedure (e.g. libertarianism), others on the outcome (e.g. utilitarianism, strict egalitarianism). Often a combination of both aspects is advocated. The most important principles, on the basis of which goods (or other things) are to be distributed, are the individual's contribution to the good which is to be distributed, the maximisation of overall social utility, equality, and need. In addition, there are different views with regard to the appropriate "currency" of justice, that is, what is to be distributed justly. Proposals include resources, welfare, capabilities, and opportunities, or concepts similar to these (cf. Sen 1979, Dworkin 1981 a, 1981 b, Arneson 1989, Cohen 1989, Daniels 1990, Page 2007 b).

The focus of this chapter is on the distribution principles under concern rather than the potential currencies of justice, as the former are more relevant to the subject here. The most important distribution principles within sustainability discourse (and also the most prominent ones in the contemporary philosophical discussion) are equality and need. One of the principles defines distributive justice by comparing different human beings or groups such as generations, whereas the other asks for an absolute standard which is to be achieved independently of what others have.[65]

Comparative and absolute standards

Comparative principles of justice always imply a claim for more equality —a principle demanding greater inequality as such, independent of additional principles such as merit, could hardly be justified[66]. Whether any

65 The philosophical debate on distributive justice features a third, intermediate position, the so-called priority view. It holds that particular attention should be given to those who are the most needy (Parfit 1997: 213). However, this view is sometimes assigned to either of the other two positions—see, for instance, Krebs (2000: 35), where Parfit's priority position is interpreted as a version of absolute standards, versus Meyer (2008: 4.2), where the priority view is seen as a derivatively egalitarian view. This view will therefore not be discussed further here.

66 Margalit (1997: 155 f.) provides an example in order to show that equality can be humiliating and unjust: he describes a number of workers who have spent very different amounts of time on the same task and in the end—in accordance with what was promised to them individually—all receive the same remuneration. However, the injustice does not arise because inequality as such is the relevant distribution principle, but because the principle of merit should be applied.

higher degree of equality in distributions[67] should be aimed for or not is subject to a comprehensive debate in philosophical literature[68].

Egalitarian theorists do not usually propose strict equality in the allocation of goods or other benefits, which would be too obviously unjust. They also do not necessarily see equality as the highest of all values, but they still see it as a crucial element of any justice theory (Gosepath 2003: 275 f., Temkin 2003 b: 769). A greater degree of equality may be subordinated in importance to other good things, for instance higher absolute welfare for the most needy, as is the case in Rawls' difference principle[69]. Arguments that distributive justice should at least to some degree be interpreted in terms of equality include in particular the following: that sometimes equal distribution follows from the norm of equal treatment (Gosepath 2003: 295, Nida-Rümelin 2006: 131); that equality as such can be intrinsically valuable (Temkin 2003 b); that undeserved inequalities, which are not based on one's own choices, are unfair (Temkin 2003 a, 2003 b, Cohen 2009); that significant inequalities can give those who are better off an unacceptable economic and political power over the lives of others (Scanlon 2003: 205); and that a certain degree of distributive equality supports community and social cohesion (Cohen 2009).

Opponents of egalitarianism criticise that it may propagate the vice of envy; that a society which is too egalitarian lacks incentives for people to work and create wealth; that egalitarians may be accused of favouring a distribution where all are poor over one where some are rich and only some are poor ("levelling down objection"; Parfit 1997); that striving for

---

67 Other application areas of equality, such as "treatment as equals", are not considered here (cf. section 2.1.2).

68 For arguments in support of equality in distributions, see for instance Scanlon (2003), Temkin (2003 a, 2003 b), Gosepath (2003), Nida-Rümelin (2006), and Cohen (2008, 2009). Many other egalitarian theorists simply assume (or have done so in the past) that distributive justice is based on some form of equality anyway, so no dedicated arguments are put forward there (e.g. Dworkin 1981 a, 1981 b, Arneson 1989, Cohen 1989, Daniels 1990, Sen 1979, 1992). For arguments against egalitarianism, see Raz (1986: 217 ff.), Frankfurt (1987, 1997), Rosenberg (1995), Parfit (1997), Anderson (1999), Krebs (2000, 2001, 2003), Crisp (2003), and Schramme (1999, 2003, 2004). For arguments both against and in favour of equality, depending on the context, see Margalit (1997).

69 This principle suggests that "social and economic inequalities are to be arranged so that they are [...] to the greatest benefit of the least advantaged [...]" (Rawls 1999 a: 266). Inequalities are therefore allowed in Rawls' theory as long as they result in a greater (absolute) benefit to the least advantaged members of society.

any greater degree of equality would be absurd after all have reached a certain level of wealth ("rich" vs. "super-rich"; Frankfurt 1987: 35, Crisp 2003); that (a certain version of) egalitarianism is inhumane, since it only requires compensating for misery which is not self-inflicted (Anderson 1999: 295 ff.); and that egalitarians are, in principle, mostly interested in the existence of a safety net and equality is only a derived value (Krebs 2000: 17 ff.)[70].

As opposed to the comparative principle of equality, an absolute standard relates to the required absolute amount of resources (or any other currency of justice[71]) that enables a person to reach a certain level of well-being. This absolute amount is often referred to as "sufficiency threshold" and the respective theorists are called "sufficientarians"[72]. In order to avoid confusion, I use the term "absolute standard", as "sufficiency" is used in a different sense here[73].

Advocates of absolute standards think that equality as such does not matter, but what matters is the absolute level of well-being which should at least reach a certain threshold. For advocates of an absolute standard, inequalities above the threshold are not of interest for considerations of justice. The fulfilment of such a standard might actually lead to a significantly higher degree of equality than before, but this is rather seen as a by-product (Krebs 2003: 240 ff.). Tricky questions arise in distribution situations where resources are not sufficient to bring all people involved to the threshold where each has enough (cf. Frankfurt 1987: 31, Crisp 2003: 758 ff., Page 2007 a: 14 f., Meyer 2008: 4.3). Advocates of absolute standards might in such cases—depending on the circumstances[74]—favour an

---

70 See also a good summary of arguments against equality and its defence in Casal (2007: 304 ff.).

71 Some non-egalitarians hold that through the adoption of an absolute conception of justice the currency debate can be avoided (e.g. Beckerman 1999: 86). However, I think it is obvious that the same question arises here as well: an absolute standard *of what*?

72 See, for instance, Frankfurt (1987), Rosenberg (1995), Crisp (2003), Casal (2007), Page (2007 a), and Meyer (2008, 2009).

73 Cf. the comparison with the concept of sufficiency as used here in section 3.2.5; cf. also Spengler (2016 a).

74 Frankfurt has outlined an extreme case where the available resource (e.g. medicine) must be distributed unequally in order to bring at least some members of a group of people to the threshold of survival, whereas all individuals would die if the resource was distributed equally (see Frankfurt 1987: 30 ff.).

egalitarian distribution, but this can also be interpreted as giving priority to the most needy with the result that equality is again achieved only as a by-product (Krebs 2003: 241, Schramme 2003: 267 f.).

An important question with regard to absolute standards is obviously where to set the threshold. The Brundlandt Commission's definition of sustainability supports an absolute standard at some minimum level where *needs* are met[75]. The term "needs", however, also requires further specification. A decades-old dispute deals with the question whether something such as needs as distinct from desires exists at all (cf. Box 1). But even if it is assumed that this is the case, still different levels of needs can be distinguished: the term may refer to the most basic needs that have to be satisfied to ensure survival, or to a somewhat higher standard, namely to provide what is required for individuals to achieve their own plan of life (Miller 1976: 134), to enable individuals to live a good life (e.g. Schramme 1999: 172), or to participate as agents in a particular society (Ware/Goodin 1990: 2).[76]

---

75 Note that some theorists have expressed doubts about whether meeting needs and humanitarian motives have anything to do with justice at all. Brian Barry, for instance, stated (1996: 205): "Humanity requires that we respond to others' needs whereas justice requires that we give them their due. If something is due to you, you do not have to show that you need it or that you will make better use of it than other possible claimants. Justice and humanity thus diverge in content." In the same vein, Thomas Nagel stated (2005: 118 f.): "Justice as ordinarily understood requires more than mere humanitarian assistance to those in desperate need, and injustice can exist without anyone being on the verge of starvation. Humanitarian duties hold in virtue of the absolute rather than the relative level of need of the people we are in a position to help. Justice, by contrast, is concerned with the relations between the conditions of different classes of people, and the causes of inequality between them." If Barry and Nagel are right and justice requires something different than meeting needs, then—considering the Brundtland definition—it might be that sustainability does not have anything to do with justice at all. There are, however, also opposing views to this (e.g. Miller 1976: 122 ff., Meyer 2009: 290). Not only do we often say in colloquial language that it is unjust if someone is in need and it is not their fault, but also the whole discussion on absolute standards is based on the assumption that justice has to do with some threshold of what is *needed* to live a decent or good life. In addition, something similar could be said for equality: what does equality have to do with giving someone their due? The relevant question is *why* is something due to someone? There are many possible answers to this. I therefore do not agree that need is not a relevant claim with regard to justice.

76 Other proposals to define the threshold not directly in terms of needs include: setting it at a level where people are content (Frankfurt 1987: 37); a level that ensures

These two points—the survival minimum and the minimum needed to live a good life—can be understood as the ends of a continuum along which the various ideas for thresholds can be allocated. The location of the upper end is certainly more difficult to determine. Most advocates of absolute standards clearly set the threshold well above the subsistence minimum (cf. Frankfurt 1987: 73, Page 2007 a: 17, Meyer 2009: 290). This is of importance in particular as they argue within a dispute in justice theory, and they do not want to see justice confused with merely satisfying a minimum that is already required for humanitarian reasons (Meyer 2009: 290).

Rosenberg (1995: 67) suggests that determining the threshold of what is enough is open to empirical investigation, and that this investigation should include the potential special needs of some individuals. Anderson (1999: 320) notes that besides individual circumstances, cultural norms and the conditions given in an environment, such as the climate, have an influence on the threshold. Other theorists, however, doubt the possibility that a threshold can be determined at all (Arneson 2006: 27):[77]

> *"A human life can be better or worse for the one who lives it. Well-being varies by degree. But it is not so clear that there is a natural way to identify a good enough level, a threshold such that it is of paramount moral importance to get people to the threshold but a matter of relative indifference what happens to them above the threshold."*

This does, however, disregard the fact that such a threshold is not only conceptionally useful, but also that in practice many institutions define thresholds, for instance with regard to poverty or social benefits (Page 2007 a: 16 f.). Unanimity is unlikely to be achieved in deliberations over the threshold, but empirical investigation and scientific evidence can be used to support the process. I will assume for the moment that it is possible to determine such a threshold (or various thresholds), and that it can be located for different purposes somewhere in between the minimum re-

---

decency, understood as the absence of humiliation (Margalit 1997); one that allows for "at least a bit more than mere physical survival" for a distinctively human life (Shue 1999: 540); one that enables individuals not to be oppressed by others and to function as equals in civil society (Anderson 1999: 320 f.); linking the threshold to situations where an external observer would feel compassion (Crisp 2003); or setting it at a level that allows human beings to develop central capabilities (Nussbaum 2011: 36 ff.).

77 Cf. also Casal (2007: 316).

quired for survival and what is needed to allow individuals to choose among a reasonable diversity of individual conceptions of the good life.

Box 1: The concept of "needs"

The concept of *"needs"* has been discussed extensively in a variety of disciplines, most notably political theory, philosophy and psychology. The idea attracts substantial attention because claims of need are used to establish entitlements or influence priorities (Frankfurt 1984: 1), including for political purposes. Referring to needs implies that they ought to be met, which results in a claim for rights (Soper 2006: 355). Or, as Rauschmayer et al. put it (2011: 10): "Needs are the most fundamental dimensions of human flourishing. We call needs those reasons for action that require no further explanation or justification."

In mainstream economic theory, the term is usually avoided in favour of another concept, namely that of preferences[78]. Preferences can be understood as comprising both needs and "wants" (or desires), although those upholding the concept of preferences would reject such a distinction, which is the very reason why they prefer to speak in terms of preferences. Economist John Kenneth Galbraith has critically remarked on this (cf. also Shue 1993: 55, Daly 1991: 40):

"[...] economic theory has managed to transfer the sense of urgency in meeting consumer need that once was felt in a world where more production meant more food for the hungry, more clothing for the cold, and more houses for the homeless to a world where increased output satisfies the craving for more elegant cars, more exotic food, more erotic clothing, more elaborate entertainment—indeed for the entire modern range of sensuous, edifying, and lethal desires." (Galbraith 1958 a: 109)

The major discussions revolving around the concept of needs are, thus, whether something like needs exists as distinct from wants, and if so, what they should comprise in practical terms.

A variety of approaches have been suggested to distinguish needs from wants. Some of them are rather descriptive, while at the same time others aim to make the case for a general priority of needs over wants. Among the descriptive ones are the following proposals: needs are (more) objective, while wants are fully subjective (Miller 1976: 129, Ware/Goodin 1990: 1); there is an epistemic difference between the two—we know what we want, but we sometimes do not know what we need (O'Neill 2011: 29); the idea of (absolute) needs is a threshold concept, while that

78 With the notable exception of John Maynard Keynes (1930), who wrote of "absolute" and "relative" needs corresponding to needs and wants.

of wants is not, so needs are satiable, while wants are insatiable (Keynes 1930, Daly 1991: 40 f., O'Neill 2011: 28 f.). The second group of proposals for distinction includes the following: meeting needs is relatively more urgent than satisfying wants, even if urgency only indicates a "rough and ready correlation" that does not always persist (Goodin 1988: 35 f.); harm (cf. section 4.2.1) is the consequence if certain needs are not fulfilled, which is not the case for unmet wants (Miller 1976: 130 ff., Frankfurt 1984: 6 f., Doyal/Gough 1991: 49 ff., O'Neill 2011; cf. also the discussion in Humphrey 2002: 56 ff.); needs indicate what is necessary to fulfil one's plan of life (Miller 1976: 133 f.), or, in other words, the "universally necessary instruments" with which to meet a wide variety of ends one might pursue (Goodin 1988: 36 ff.).

All of these characterisations can be and have been challenged as either too vague, paternalistic, or not enduring when confronted with certain examples. A problem in the last sense is the case of instrumental needs that are trivial. It is argued (against the definition by Rauschmayer et al. quoted above) that needs are always instrumental in that what is needed is required for some purpose, even if it is not always necessary to mention this purpose because it is obvious (Ware/Goodin 1990: 1). Thus, we often also refer to the term "needs" when we mean that something is necessary for a trivial purpose, for instance in Harry Frankfurt's example of a crossword puzzle: Someone wants to complete a crossword puzzle, but is unable to do so without using a dictionary. This person therefore *needs* a dictionary; however, "the moral importance of this need is altogether negligible" (Frankfurt 1984: 2).

Frankfurt has therefore attempted to distinguish between different types of needs in order to identify those of particular moral urgency: When something is needed because there is some end that a person *wants* to achieve, then the need depends on the person's will and is therefore "volitional". Free volitional needs, according to Frankfurt, are not morally interesting, because no harm will ensue if they are not met (Frankfurt 1984: 4 ff.). However, Frankfurt's account remains problematic if the reason for distinguishing genuine needs from wants is that one intends to promote the thesis that (certain) needs are generally to be prioritised over wants. This would imply that the ends served by "non-volitional" needs should be prioritised systematically over those ends served by free volitional needs (or in other words: wants) independently of the bearer's own prioritisation (Goodin 1988: 30 f.), which would be outrightly paternalistic (Goodin 1990: 13, cf. also Soper 2006).

Still, many authors insist that, even though the distinction between needs and wants and the consequences that follow from this distinction remain highly contested, there simply is such a thing as basic needs as opposed to purely trivial wants (e.g. Daly 1991: 41, Shue 1993: 55, Sanne 2002:

283, cf. also Goodin 1990: 26, Soper 2006: 361). In line with these authors, even though a clear distinction may not be possible, I think this is no reason to reject the distinction as such, and that there are good reasons to prioritise the satisfaction of basic needs over the satisfaction of certain wants.
A second question, then, is what needs should comprise in practical terms. Some theorists have proposed concrete lists or categorisations of needs (e.g. Doyal/Gough 1991, Max-Neef et al. 1991, Hamilton 2003). The Brundlandt Report is concerned particularly with "basic" or "essential" needs of the world's poor, the fulfilment of which should have "overriding priority" (e.g. WCED 1987: I/2/1, 4, 29, 42 ff.). The report, however, does not specify these in detail; it only indicates by means of examples that they comprise primarily basic physical needs, such as water, sanitation, food, clothing, shelter, as well as jobs (Rauschmayer et al. 2011: 12). What is more problematic than some rough agreement about basic needs are attempts to define non-basic needs, or needs that are not absolutely required for human survival but have become a common standard in industrialised countries, such as cold and hot piped water at home (cf. Goodin 1990, Jackson 2005, Soper 2006). Much discussion within consumption research (cf. section 3.2.5) therefore revolves around needs that are, even if not required for human survival, conditional for the realisation of a diversity of conceptions of the good life, and can be somehow distinguished from unneeded "luxury" goods.

After having outlined the two most relevant current positions within the debate on distributive justice, egalitarian and absolute standards, I now turn to the specific challenges related to international and intergenerational justice. Further below, I will present what follows from these with regard to the discussion on egalitarian and absolute standards, and a pragmatic position will be adopted in light of the sustainability debate and the questions raised in this study.

Particular challenges of global justice

The crucial question around which discussions about global justice revolve is whether any difference should be made in justice relationships between people in the same state (or other group, such as a nation or people), and between them and people in other states (or other groups). Such a potential difference is decisive for the question of whether, and to what extent, redistribution of economic assets across state (or group) borders should take place.

Many theorists suggest that, indeed, not the same demands do apply to justice internationally as within a state, and provide different justifications for this. One potential justification, held by those who emphasise the importance of community affiliation, is simply that national identity matters to people and should therefore be seen as morally relevant in framing obligations of justice. For instance, Miller has argued that while intense, egalitarian demands of justice apply at the national level, this is not the case internationally: on the global level there is no "shared identity", no common "ethos", and no similar institutional structure that acts on behalf of the community; therefore, a hypothetical world community is not suitable as a ground of justice (e.g. Miller 1999b: 189f.). Such a nationalist account is, however, hardly acceptable for most liberals, and Miller's arguments have been criticised as inconsistent and implausible (e.g. Caney 2003, Weinstock 2003). For example, it has been argued that if, following Miller, the demands of justice depend on group membership, then the relevant group should be defined by its interdependence through a common socio-economic order and not by something as arbitrary as national boundaries (Caney 2003: 295).[79]

Another, much more widespread way to justify differences between national and international justice relationships deals with the disparities in the national and international orders, or, in other words, asks whether some "normative peculiarity of the state" exists (Risse 2012). Many theorists refer here to the coercive nature of the state as opposed to the anarchic international order (e.g. Nagel 2005, Risse 2012). It has also been stated that basic decisions with regard to the national economic order are taken within states, and that respecting these decisions requires not calling for extensive redistribution across state borders (Miller 1999b: 193f.) or between peoples, respectively (Rawls 1999b: 117f.). Risse mentions two features of states that form the basis for special justice relations within a state: not only the power to enforce rules and coerce immediately, "without having to go to war", but also some specific sort of cooperation between free and equal citizens that he calls, following Rawls, "reciprocity" (Risse 2012: 24, cf. Rawls 1999b: 14). These theorists deny that the inter-

79 This alternative, which Caney mentions, actually reflects the position of other theorists. Caney himself goes even further than this, claiming that even without such a link the same requirements of justice may exist.

national order can be interpreted in a similar way as a substantial cooperative and coercive scheme.

Others have, however, objected to this. An increasing number of justice theorists point to the empirical circumstances of international justice relations and the actual reasons for extreme poverty in poor countries. Not only do they highlight the fact that in the more recent past and today, there has been and is a substantial amount of economic and other cooperation between countries, based on which rules relevant for distributive justice are established (e.g. Caney 2003: 296, Satz 2003: 626 f., Cohen/Sabel 2009: 164 ff.); some also hold that the existing system of international cooperation is in part causally responsible for the misery in some places of the world and that people in rich countries benefit from this system (Pogge 2004, 2005, Young 2006). Pogge especially has defended the view that it is wrong to think of transfers from rich to poor countries only in terms of voluntary aid or "assistance", but rather that from the given global order that benefits the rich, strong negative duties arise on the part of affluent people and countries to stop harming the poor and compensate for the harm caused (Pogge 2004, 2005; cf. also Young 2006).

Particular challenges of intergenerational justice

In comparison to the intragenerational debate, including future generations in the "scope" of justice involves a different set of challenges.

Most notably, in intergenerational justice relations future generations are "downstream" of present ones, so it is impossible for them to influence the decisions made in the present, but they are affected by these decisions (Rawls 1972: 291). Present actions do, for instance, have an influence on the state of the environment in the future and on population size. Meyer (2008) characterises this as a "permanent asymmetry in power-relation" between present and future individuals.

Further, uncertainty about the future generally exists. We do not know what the effects of human activities on the environment will exactly be in the long run. We also do not know what benefits science and technology may bring in the future. Over the last few hundred years and in industrialised countries at least, later generations were generally better off than earlier ones in the long term (with some exceptions, such as in times of war), if we consider the most important welfare indicators. However, from past developments it cannot be concluded that this trend will continue forever. In addition, the potential consequences of technologies such as nuclear power pose substantial risks to future well-being. Therefore, we cannot treat existing and future generations fully alike because our predictions for the future are characterised by uncertainty (cf. Box 2 in section 2.3.4). What is more, the number of people living in the future among whom the goods or resources available by then will need to be distributed is, at least for the more distant future, not known (Barry 1999: 107 f.).

Related to this is the discussion on the discounting of future goods in comparison to present ones. Discounting is a common practice in economics that is done, among other reasons, because of uncertainty and the preference of having something now rather than later ("time preference"). It is also in line with interest rates paid for the alternative of depositing money in a bank instead of spending it on a good, so that today less money is needed for the same good compared to the future. While widely accepted for shorter time periods, discounting is controversial in the intergenerational context (Birnbacher 1977: 399, 2001, Pearce/Turner 1990: 211 ff., Krebs 2001: 164, Lerch/Nutzinger 2002, Ott/Döring 2004: 120 ff.). Discounting calculations lead, depending on the discount rate used, to extremely low values attached to even fundamental goods in the long term (e.g. 100 years). It is therefore argued that mere time preference cannot be a morally valid argument to discount, for instance, human life in 500 years compared to human life today (Krebs 2001: 164).

Another question related to intergenerational justice is the time horizon: how far should justice considerations reach into the future (cf. Pearce 1988: 598)? The Brundlandt Commission's definition does not contain any limitation in time. Sustaining something means that in principle a certain state should be able to last indefinitely and that no inherent factors undermine this possibility (irrespective of external and unavoidable influences such as the end of our solar system in the very distant future).

Finally, a (rather academic) problem concerns the identity of individuals in the future ("non-identity problem"; Parfit 1982). It has been pointed out that any decision taken today, including those that are supposed to benefit generations to come, affects the actual existence and identity of people in the future, with the result that people with a different identity will be born than would have been the case without, for instance, a certain political measure. In consequence, the question is raised whether it would be better to be born under certain (awkward) circumstances than not to be born at all. It has been suggested in this context that in the vast majority of cases it is better to be born than not to come into existence at all. This has given rise to the paradox that future individuals cannot have rights today, since any action on their behalf would probably impede them being conceived in the first place. This problem has been discussed extensively elsewhere, and convincing arguments have been proposed against the denial of present duties towards future generations due to the uncertainty of people's identity.[80] If we can expect that human beings will exist in the not too distant future, which is very likely, it does not matter how many and who exactly these individuals will be with regard to the general ability of people in the present to harm the rights of future ones.

All in all, justice towards future generations remains more abstract than justice relations between people who are currently alive, as we cannot now distribute concrete assets to future individuals (on a society-wide level). It is impossible to ensure now that a certain kind and level of distribution will be achieved in the future, since circumstances can always change unexpectedly. What can be done now is just to save an overall amount of resources and work towards the persistence of certain framework conditions, so that the likelihood of a certain standard of justice being met in the future is increased.

A just distribution pattern internationally and towards the future

What is the consequence of these special problems of global and intergenerational justice? Basically, some theorists use them to explain why they think different standards of justice are due to non-compatriots or future generations compared to contemporary compatriots. Others, however, challenge the correctness of the assumptions made by the former or deny that these differences should play a role at all in considerations about justice. Besides some extreme positions, by and large both the debates on global and intergenerational justice reflect the discussion on absolute and comparative standards.

Within the debate on global justice, for those who claim that the (nation) state has special importance for demands on justice this means that less stringent, non-egalitarian demands apply outside the state. Most theorists who hold this view do not say that no justice requirements apply internationally. Some suggest, for instance, that rich, well-ordered states

80 Summaries e.g. in Meyer (2008, 2009: 293 ff.) and Tremmel (2008: 34 ff).

should support other states in becoming well-ordered and respecting basic human rights (Rawls 1999 b: 106-119, Miller 1999 b: 197 ff., 2004 b, cf. also Blake/Smith 2013). Such demands are not comparative: they rather require wealthy states to help others to meet a certain minimum baseline. According to Miller, a cross-cultural moral minimum exists which "reflects the conditions that are universally necessary for human beings to lead minimally adequate lives", and if this minimum is met, that is, in the absence of real deprivation, inequalities between two states do not matter morally (Miller 1999 b: 189, 199).

Others argue that just ensuring that basic needs are met in other countries does not suffice for wealthy states to satisfy their duties, given the actual global interdependencies and the beneficial effects of the current global order for most wealthy states. They therefore suggest that rich countries owe poor ones more than just an absolute humanitarian standard (Cohen/Sabel 2009: 164), or even that international egalitarian principles of distributive justice arise from this (Pogge 1994: 196) or exist independently of any interdependencies (Caney 2003: 296 f., Singer 2004).

Likewise, there is disagreement within the intergenerational debate about the just distribution pattern between people in the present and in the future. The views range from doubts about whether present generations have any obligation towards future generations at all, to the opinion that within sustainability debates justice towards future generations should be given priority (Kopfmüller et al. 2001: 142, Nutzinger 2006: 194, Ott/Döring 2004: 73 ff.).

The first extreme, the "no obligation view", is based on the non-identity problem. It is rejected by most participants in the debate (see e.g. Krebs 2001, Meyer 2008, Ott/Döring 2004: 73, Tremmel 2008: 56 ff.). It is stated that in order to justify that there are obligations towards the future, it is sufficient to acknowledge that people in the future will have rights once they are there (SRU 2002: 58). As the other extreme, prioritising future generations within a conception of sustainability could be considered as distribution among contemporaries is already a subject of other theories (Kopfmüller et al. 2001: 142). However, it has been stated that it is untenable to ignore present poverty while focusing on the well-being of future generations (SRU 2002: 58, Christen/Schmidt 2011: 5). This is also not intended by the Brundlandt Commission's sustainability definition.

The two main intermediate positions are to concede an egalitarian or an absolute standard to future generations. An egalitarian standard for future generations means that they should not be worse off in terms of well-being, opportunity, resources etc. than present generations. "Not worse off" implies that people in the future could also enjoy a higher standard of well-being than earlier ones, because freezing well-being at the present level regardless of possible unexpected developments would not make sense.

It is necessary to clarify what such an egalitarian standard for future generations would imply. Among present individuals, it is possible to share, say, a cake in such a way that, in the end, we can say for sure who has received how much. This is, however, not possible with regard to the future: at the moment, we can only save a certain part of the cake for the future, but we do not know how many people will exist then and how they will decide to share what has been saved among them. Thus, we cannot say how much each will get exactly, and so intergenerational egalitarianism should rather be thought of as a doctrine which requires the maintenance of the overall amount of resources or, to use a term which is common in the sustainability debate, capital (cf. section 2.3.4), for the benefit of whole generations, that is, the *sum* of people who exist or will exist at a particular point in time.

Some of the arguments in favour of an egalitarian distribution among contemporaries do not hold across generations—for instance, the arguments for greater social cohesion and the avoidance of undue exercise of power by the rich over the poor (Beckerman 1999: 80). However, other egalitarian arguments based on equal treatment, a possible intrinsic value of equality, and the idea that undeserved inequalities are unfair, do apply in the intergenerational context as well. The view of intergenerational egalitarianism is therefore intuitively convincing, as there is no reason to assume that any generation should be anything special in the chain of generations over time (Ott/Döring 2004: 95 f.). However, as Beckerman (1999: 73) points out, it is also difficult to see why exactly the *current* average level of welfare should set the standard below which future generations should not fall. Beckerman, and also other theorists (e.g. Meyer 2008, 2009; cf. also Page 2007 a), thus propose an absolute standard for intergenerational justice. Rawls' "Just Savings" principle is also often interpreted in this sense (Rawls 1972: 284; cf. Schramme 2006: 154, Meyer 2008)[81].

81 For a somewhat different interpretation, see Wall (2003).

Synthesis

Having reviewed, at least to the extent possible within this study, the major current debates on distributive justice within states as well as globally and between generations, I suggest that the majority of positions with regard to distributive justice can be brought under a common denominator[82]: They all support the idea that *at least* conditions should be created and safeguarded that enable everyone to meet their basic needs and avoid being harmed and, in Henry Shue's words, "at least a bit more" than this, "for at least a distinctively human, if modest, life" (Shue 1999: 540 f.)[83]. This would correspond to an absolute standard located somewhere at the lower end of the aforementioned continuum, and can thus be called a conception of "minimal justice", which could form the basis of further-reaching conceptions of justice (cf. Nussbaum 2011: 40).

Such a minimal standard is in line with sustainability's primary aim of avoiding human misery now and in the future by safeguarding conditions that allow for meeting human needs. In the eyes of many of those involved in the justice debate, it would, if realised, not lead to a just world, but at least to a "juster world" (Lerch 2006: 122). The dispute about whether we should in addition strive for any greater degree of equality seems to be of comparably lower importance given the present proportions of worldwide poverty and suffering, as well as the doubts about whether future generations will be able to satisfy their needs in view of the risks of climate change and other environmental hazards.

2.3.4 Weak and strong sustainability—the role of natural capital

In addition to the question of which distribution pattern would be just in the sustainability context, with regard to future generations the question of *what* should be sustained so that they can meet their needs is controversial.

---

82 Except for libertarianism or arguments based on the intrinsic value of equality. Cf. White's (2008) analysis of the views of different theories (utilitarianism, libertarianism, egalitarian liberalism etc.) with regard to the enactment of a "social minimum".

83 Cf. also White (2008): "We define a 'social minimum' as that bundle of resources which suffices in the circumstances of a given society to enable someone to lead a minimally decent life."

Different proposals have been made in this context, ranging from the extremes of "very weak" to "absurdly strong" sustainability. The "strength" of each of these conceptions is mainly determined by the permitted degree of exchangeability or substitutability[84] of different types of capital, in particular between natural capital on the one hand and man-made capital on the other. The former can further be split up into natural and cultivated natural capital (or even into more detailed subcategories), the latter into knowledge, social (e.g. institutions), human (education, health), and man-made material capital (Ott/Döring 2004: 100, Michelsen/Adomßent 2014: 32). The term "capital"[85] refers to "a stock of materials or information that exists at a point in time" and each form of stock

> *"generates, either autonomously or in conjunction with services from other capital stocks, a flow of services that may be used to transform materials, or the spatial configuration of materials, to enhance the welfare of humans. The human use of this flow of services may or may not leave the original capital stock intact"* (Costanza et al. 1997: 254).

The common denominator of these different sustainability conceptions is the condition that the sum of all capital bequeathed to later generations—or the utility[86] derived from that capital—must not be lower than before, and thus all are in principle based on egalitarian reasoning. Accordingly,

---

84 The issue of substitutability is clearly at the core of the debate on weak vs. strong sustainability, but sometimes also other controversies are subsumed under this headline, for instance the discounting problem (see section 2.3.3) or the question of anthropocentrism vs. physiocentrism (section 2.3.2).

85 The use of the term "capital" for natural assets is common within the discussion on weak vs. strong sustainability. Similar and related terms, such as "natural resources", environmental "sinks" and "sources", or "environmental carrying capacity" are used within sustainability discourse, too, and I use all of them somewhere within this text. As "natural capital" has a strong economic connotation which many dislike, I restrict the use of this term mainly to this section, and elsewhere will mostly use the term "environmental carrying capacity" (on its relation to natural capital, see further below in this section). It should further be noted that there is no agreement on a specific definition of natural capital (cf. Pearce/Turner 1990, Neumayer 1999: 9, De Groot et al. 2003: 188, Brand 2009: 608; see also the discussion in Ott/Döring 2004: 194 ff.). As an alternative to a definition, lists are proposed that contain elements that obviously belong to natural capital, such as the atmosphere, the climate system, soil, rivers, fish stocks, forests etc. (cf. SRU 2002: 65).

86 The term *utility* is mainly used with regard to the paradigm of weak sustainability, as it is needed there to compare the value assigned to different capital types.

some use the definition that "development [...] is sustainable if it does not decrease the capacity to provide non-declining per capita utility for infinity" (Neumayer 1999: 9).

The main point of controversy is the possibility of substituting natural capital with man-made capital. The position of weak sustainability holds that it is sufficient to ensure that the total stock of capital (or the derived utility) is maintained at the same level, so that in principle it is not necessary to preserve a defined share of the stock of natural capital. Within this paradigm, natural capital can be substituted by other capital types through technical progress as long as the total capital, or the utility derived from it, does not decline. It is assumed that the preferences of future generations will adapt continuously to a changing environment, where more and more natural capital is replaced by man-made capital (Ott/Döring 2004: 103 f.).[87] In the vision of "very weak" sustainability, a theoretical situation in which *all* natural capital would be replaced by man-made capital of the same value, or be able to bring about the same level of utility, would still be called sustainable.[88] That this would be the case is, however, very unlikely, because, as advocates of weak sustainability stress, even within this paradigm it is impossible for natural capital to decline until below certain levels because this would have too strong drawbacks on overall utility, which could not be compensated for.

However, the substitution debate, which is held in very abstract terms, gives rise to terminological issues if thought through more carefully. A clear distinction between natural and man-made capital does not seem possible. Natural capital is often needed to produce man-made capital (Ekins et al 2003: 169, Nutzinger 2006: 198 f., Pearce/Turner 1990: 48 f.), at least the types of man-made capital which are not immaterial. For example, mankind will always need food to survive, and even the type of

---

87 But, as Ott and Döring note, the opposite could just as well be the case, so that individuals would develop strong preferences for natural sites in a world where these have become scarce (2004: 104).

88 A famous quotation from Robert M. Solow is often interpreted in this sense: "If it is very easy to substitute other factors for natural resources, then there is in principle no 'problem'. The world can, in effect, get along without natural resources, so exhaustion is just an event, not a catastrophe" (Solow 1974: 11). However, Solow is referring only to exhaustible natural resources, which is an important point, as this allows for the substitution of non-renewable with renewable resources. This possibility is not called into question by strong sustainability (see below), in contrast to the possibility of substituting *all* natural capital with man-made capital.

food that is intensely processed and seems "unnatural" is made up of natural resources. It is therefore unclear what it would mean if, in the extreme case of "very weak" sustainability, all natural capital was replaced by man-made capital. On the other hand, the possible conclusion that all material things constitute natural capital would not be very helpful either. It would therefore make more sense not to refer to the physical matter contained in capital but to the services provided by different capital types (cf. Costanza et al. 1997) in the distinction between natural and man-made capital.

The position of strong sustainability regards natural capital generally as not substitutable by man-made capital or only substitutable by it to a very limited extent. According to Daly, man-made and natural capital are complementary in creating utility (Daly 1996: 76 ff.), so that the share of one of them may not be increased at the expense of the other (SRU 2002: 64). If man-made and natural capital are complements, one of them—the one shortest in supply—will be a limiting factor, which nowadays is supposedly more and more often natural capital (Daly 1996: 78, Goodland/Daly 1996: 1005). However, the complementarity thesis established by Daly has so far not been backed sufficiently by theoretical considerations, but relies mainly on intuitive examples, such as a sawmill without a forest or a fishing boat without fish (cf. Daly 1996: 76 ff., Ott/Döring 2004: 139 ff.). It has thus been stated that complementarity versus substitutability is a rather empirical question (Lerch/Nutzinger 2002: 8), and accordingly research has been carried out to find the "limits to substitution" (Ayres 2007, Ekins 2003, Ekins et al. 2003). Further arguments for strong sustainability include the uncertainty about the consequences of natural capital loss, its irreversibility, the fundamental relevance of the functions natural capital provides for human beings, the greater freedom of choice for future generations (Neumayer 1999: 27, Ott/Döring 2004: 160 f.) as well as the normative question of whether we "should want" to substitute nature with artificial goods, regardless of its technical feasibility (Ott/Döring 2004: 144, see also Tribe 1980).

Strong sustainability therefore establishes a requirement in addition to the claim that the total stock of capital must be maintained, which also holds that the stock of natural capital as a sub-amount of the total capital stock must be kept constant (Daly 1996: 76, Neumayer 1999: 27; cf. Pearce 1988, 1990). The position of "absurdly strong" sustainability would not allow any subtype of natural capital to be diminished, so no substitution with other types would be possible. This means that, for instance, depletion of fossil fuel resources would not be allowed at all as they are non-renewable, so that less would be left for later generations, and renewable resources could only be used to the extent produced by net growth rates (Goodland/Daly 1996: 1006). Since such a rule would neither benefit present nor future generations (provided that the rule lasted into the future), it is considered "absurd". Strong sustainability thus allows for the exploitation of non-renewable natural capital as long as substitution of its functions with renewable natural capital takes place. Strong sustainability, therefore, accepts some substitution *within* the class of natural capital (Ott/Döring 2004: 147).

What leaving the total amount of natural capital intact would mean requires a more precise definition, however. Just measuring the quantity of physical capital stock would result in a rather useless aggregate number of all types of natural capital. Alternatively, it could be referred to its value, the value of its services, or constant prices (Pearce 1988: 604 f., Pearce/Turner 1990: 52 f.). Most often, the requirements of strong sustainability are defined by means of a set of "management rules". Different versions exist, but a core of three fundamental rules that focus on "sources" and "sinks" can be identified[89]:

- The depletion rate of renewable resources must not exceed their regeneration rate.
- The depletion of non-renewable resources[90] is to be minimised and must not exceed the rate at which renewable substitutes are developed,

---

89 These three rules are, besides other rules, already contained in the Brundlandt report (cf. WCED 1987: I/2/4-15); see also Pearce and Turner (1990: 44 f.), Enquete-Kommission (1994: 29 ff.), Goodland and Daly (1996: 1008), Huber (2000: 4 ff.), Kopfmüller et al. (2001: 63), Lerch and Nutzinger (2002: 6 f.), SRU (2002: 67), Ekins et al. (2003: 175), Ott and Döring (2004: 162). Additional rules sometimes mentioned state, for instance, that population size must be within the planetary carrying capacity (Huber 2000: 4, WCED 1987: I/2/7), that extinction of species should be avoided (WCED 1987: I/2/13), or that the pace of anthropogenic changes must be in the order of recovery rates through natural processes (Enquete-Kommission 1994: 32 f). These only make some special cases that are already contained in the aforementioned core of rules more explicit: for instance, species can be seen as a non-renewable, non-substitutable resource, which thus may not be "used", that is, extinguished.

90 Note that the exploitation of non-renewable resources is, strictly speaking, not an environmental issue as far as only the displacement of the resource itself is concerned (and not the often highly toxic emissions caused during the extraction process or the environmental impacts linked to the further processing of the resource;

recycling technologies are generated, or resource productivity is increased.

- Pollutant and waste emissions must not exceed the absorption and regeneration capacity of ecosystems (and emissions of persistent pollutants are to be minimised).

These management rules can be used to guide sustainability policies, for instance in the development of quantified environmental targets.[91]

However, it is difficult to understand why there should be no substitution of natural with man-made capital within a conception of strong sustainability. Given that in the past a substantial amount of natural capital has already been converted into man-made capital without compensation, it is unclear why the status quo should be the point at which no more natural capital conversion should be allowed[92]. The only plausible reason, in my view, would be if empirical evidence showed that we have coincidentially reached a point at this moment where any further reduction in natural capital would pose a significant problem of some sort. However, it might well be the case that we already reached this point some time ago, so that actually sustainability should call for an *increase* in the amount of natural capital instead of keeping it intact. This again raises the issue of a possible absolute threshold (cf. section 2.3.3). I will therefore focus in the following on a different sustainability conception, which incorporates such a threshold.

cf. chapter 1), because fossil fuels and ores are usually not relevant to the ecosystems surrounding them. However, the depletion of non-renewable resources is, of course, an issue for sustainability as it is relevant for distributive justice.

91 Huber's (2000: 271) statement that the rules are "empirically empty categorical imperatives" is not correct because of the impossibility of defining clear limits for population size and environmental carrying capacity. Nobody claims that the rules should be followed by scientifically precise and correct quantitative figures; in the light of uncertainty, which always exists with regard to environmental questions, the rules can, however, serve as guidelines. Cf. also the evaluation of the environmental status quo against these rules in Meadows et al. (2004: 55 ff.).

92 More or less the same argument has been raised with regard to intergenerational justice (cf. section 2.3.3): there, it is also unclear why, corresponding to an egalitarian position, exactly the *current* level of welfare should be maintained.

"Critical" sustainability

This conception focuses on "critical" stocks of natural capital that may not be depleted so that we safeguard basic functions of natural capital that are essential to human life (cf. Pearce 1993: 16, De Groot et al. 2003, Ekins 2003, Ekins et al. 2003). It can be traced back to a proposal for a "safe minimum standard of conservation" from the middle of the last century made by resource economist Siegfried von Ciriacy-Wantrup (1952). The proponents of this intermediate position between weak and strong sustainability consider natural capital as partially complementary and partially substitutable with man-made capital (Ekins et al. 2003, Lerch/Nutzinger 2002: 5, Ott/Döring 2004: 148, Pearce 1997)[93]. While the complementary part of natural capital constitutes the minimum "critical" natural capital stocks, the substitutable remainder consists in natural capital exceeding those critical stocks. Ehrlich and Goulder use the similar notion of "essentiality" instead of criticality, which they place in contrast to non-substitutability: "Essentiality of a given natural resource does not rule out its substitutability. Rather, it means that substitution is not possible once the stock of that resource falls below some critical value" (Ehrlich/Goulder 2007: 1150).[94] The same idea of a threshold is expressed in the term "environmental carrying capacity", which denotes the capacity of the environment to assimilate emissions to environmental sinks and to compensate for human extraction of natural resources up to a point beyond which environmental systems would be disturbed significantly and there would be harm to human beings[95].

93 Sometimes this sustainability conception is actually interpreted as a representative of strong sustainability (e.g. Pearce 1997); however, while these two might indeed tend to converge in practice, I think there is a significant difference between referring to a critical threshold and referring to equality, and it is a great achievement to frame this criticality-based conception of sustainability in much clearer terms than the vague strong sustainability requirement of maintaining something over time.

94 A certain stock of natural capital would therefore become scarce in an absolute sense when it falls below that critical level. On the notion of (absolute) scarcity, see Box 7 in section 4.3.4.

95 As opposed to my use here, cf. the use of the term carrying capacity by Arrow et al. (1995: 520 f.), who frame it as a natural limit to the scale of the economy and human population, which is non-static and "contingent on technology, preferences, and the structure of production and consumption".

As stated in the introduction to this chapter on sustainability (section 2.3.1), it has been claimed that sustainability can be established as an "objective concept" based on scientific findings about environmental carrying capacity, rejecting the "social constructs" of weak, strong and similar versions (Hueting/Reijnders 1998: 140 ff.). Sustainability, on this account, is understood as requiring the preservation of environmental functions and "[t]he 'admissible' environmental burden that goes with the definition of sustainability in terms of conservation of the functions of environmental capital", and "can, in principle, be established scientifically [...]" (Hueting/Reijnders 1998: 142, 145). The conception of critical sustainability resembles such an objective approach; however, several authors have expressed doubts with regard to this objectivity (e.g. Kopfmüller et al. 2001: 126 ff., Ott/Döring 2004: 148, Beemsterboer/Kemp 2016: 72). Not only is what should be included in critical natural capital controversial, but so is the level at which thresholds should be fixed.

As regards the first point, a research project that aimed to develop a classification for critical natural capital came to the conclusion that critical natural capital cannot simply be identified as particular elements of natural capital (cf. Ekins et al. 2003, Ekins 2003). As different natural capital elements interact in a complex way, the characteristics of the natural capital stock that enable these *interactions* need to be maintained. The resulting environmental functions that are important for us rely as much on such interactions as on the natural capital itself (Ekins 2003: 277 f., cf. also Ott/Döring 2004: 176). Environmental functions, according to Ekins et al. (2003: 175), are categorised into source and sink functions (the environment's capacity to provide resources and absorb pollutants), life-support functions ("the capacity to sustain ecosystem health and function"), and other human health and welfare functions ("the capacity to maintain human health and generate human welfare in other ways"). Determining what belongs to critical natural capital involves a declining level of certainty: for some elements, for example, the ozone layer, it is much more obvious that they are essential than for others, such as a particular species of beetle (Ott/Döring 2004: 209). Reversely, caution is needed if certain environmental functions or elements of natural capital, such as a particular species, should be labelled as non-critical. This could make a function or element "fair game", although the potential effects of its loss are not known with certainty (Ekins et al. 2003: 174 f.). Thus, while there should be agreement, based on the current state of scientific understanding, that certain elements and functions of natural capital are objectively critical and necessary for human beings, disagreement will arise when it comes to the details. The farther away one moves from the more certain end of the scale, the more likely one is to meet with normative considerations about the need to maintain natural capital.

As far as the second point is concerned, there is also a great degree of uncertainty with regard to the critical thresholds for many environmental functions (cf. Box 2). It is very difficult to determine how much natural capital stock can be used up before it ceases to deliver a certain environmental function, and searching for this point empirically would be too risky as some environmental changes are dangerous and irreversible (Pearce 1997). This is due to the complexity of the interactions between different natural capital types and functions. As a consequence, precaution (cf. Box 2) would require defining the amount of critical natural capital with considerable safety margins. It may also very well be the case that for some capital types, the amount of critical capital has already been reduced to a level lower than the critical level, so restoring it to higher levels may be required (Ekins et al. 2003: 184). The level at which such thresholds should be fixed is also related to normative considerations. This is not only true with regard to judgements related to risk and uncertainty (Rockström et al. 2009 b), but also as regards the role of the threshold: At least for some environmental functions and natural capital types, a difference can be made between a minimum level that is necessary for the survival of humanity and a level that enables human beings to have a good life (cf. Ott/Döring 2004: 176). The former is an absolute minimum requirement, as it deals with the avoidance of harm to human beings globally and in the future. Such considerations are again linked to the conception of justice used (cf. section 2.3.3). With a view to the "minimal" conception of justice used here, it follows that at least a level of critical natural capital that allows human beings to meet their basic needs should be safeguarded.

Planetary boundaries

An approach which can be interpreted as an attempt to identify and quantify such a minimum level of critical natural capital has been suggested in the recent work of an international research group on "planetary boundaries" (Rockström et al. 2009 a, 2009 b, Steffen et al. 2015). The boundaries identified must not be transgressed (permanently) in order to define a "safe operating space for humanity". Crossing these boundaries could generate “unacceptable” environmental change, which is characterised as potentially irreversible and in some cases abrupt, thereby posing a threat to humanity as a whole. These possible changes might also destabilise important systems that have kept the Earth in the stable situation it has been for the past 10,000 years, known as Holocene, and in which it would, without human intervention, continue to be for a long time (Rockström et al. 2009 b).

The proposed boundaries are in part of global character, driving important systemic processes on a planetary scale, and in part more of regional or even local character and become a global problem only when effects from many local sources accumulate (Rockström et al. 2009 a). The boundaries concern climate change, ocean acidification, stratospheric ozone depletion, the nitrogen and phosphorus cycles, global freshwater use, land-system change, biosphere integrity (functional and genetic diversity), atmospheric aerosol loading, and so-called "novel entities" (substances that have the potential to create unwanted geophysical and/or biological effects[96]). In four fields—climate change, the nitrogen and phosphorus cycles, land-system change, and genetic diversity loss—the boundaries have already been transgressed, reaching into a zone of increased risk and uncertainty (cf. Box 2) about the potential effects, or even beyond that (Steffen et al. 2015). Two boundaries, atmospheric aerosol loading and "novel entities", as well as the part of biosphere integrity that deals with functional diversity, have not yet been quantified.

The selection of boundaries shows that the proposal is not complete in relation to critical natural capital. It has a strong focus on avoiding catastrophic global risks and concentrates on emissions and pollution, that is, the sink functions of the environment as well as on several systemic life-support functions. The source functions, that is, the availability of natural resources, are only considered to a limited extent, even if they are implicitly covered in some boundaries[97]. The same is true for what Ekins et al. have called “other human health and welfare functions” (see above), for example special landscapes and space for recreation. While it can be dis-

---

96 In the first publication on the planetary boundaries, this boundary was called “chemical pollution” (cf. Rockström et al. 2009 a, 2009 b).

97 The only boundary that explicitly points out a source function is that of freshwater use. Implicitly, the source function is also addressed in biosphere integrity (although the boundary refers mainly to the extinction of species and not their abundance) and land-system change.

cussed whether these functions are really critical in terms of a minimum requirement for meeting basic needs, they are definitely important if criticality is set at a level enabling human beings to live a good life.

A more complete approach to identifying critical natural capital would therefore also require pointing out limits to resource use and possibly also to the purpose of safeguarding environmental welfare functions.

Box 2: Risk, uncertainty, and precaution—definitions and debates

The question of how to deal with risk and uncertainty is crucial for sustainability, as the concept is preoccupied with long-term developments. Some environmentally relevant activities only or still unfold their harmful effects after decades, and often the causal chains are so complex that it is hard to predict their potential—sometimes irreversible—effects. Social and economic developments are even more difficult to predict, and are interlinked with the environmental sphere. As opposed to the term "hazard", the concept of risk points to the possibility of influencing an outcome by means of targeted action, that is, defining something as a risk implies that according action may be taken to avoid an undesirable outcome or at least mitigate its impacts (Renn 2008: 1, Heinrichs 2013 a: 221).

A *risk* is determined by the extent of potential harm and the probability of its occurrence. Risks are thus clearly defined in that both the potential outcomes as well as their respective probabilities are known. With regard to the consequences of running down a specific part of natural capital or not, this is rather seldom the case (Neumayer 1999: 99). Shrader-Frechette (1991: 5, 1998: 331) distinguishes between individual risks, such as those linked to certain diets, and societal risks, such as those emanating from large-scale technologies. While the former are individually chosen, the latter are often involuntarily imposed, which is why they tend to involve more ethical controversy. Environmental risks are usually societal risks.

Uncertainty describes the possibility of potential errors in estimating a risk, as well as cases of risk for which probabilities, or the extent of the potential consequences, or both cannot be quantified (Birnbacher 1988: 141 ff., Shrader-Frechette 1991: 101 f., Wynne 1992: 114 ff., WBGU 1998: 38, Neumayer 1999: 98 ff., Lin 2006: 968). For risks which are characterised by uncertainty, at least ranges of probabilities and harm can often be estimated. A special category of uncertainty, also called "ignorance", describes cases where it can be reasonably assumed that significant further, unknown consequences of unknown probabilities may occur ("Don't know what we don't know"; Wynne 1992: 114). An example of ignorance is the problem of biodiversity loss, as it is impossible to assess the potential consequences of the disappearance of a significant share of the millions of existing species (Neumayer 1999: 102). While cases of ignorance are specifically problematic, at least even here maximal possible harms can often be roughly estimated (Birnbacher 1988: 152 f.).

Proposals and strategies on sustainability often include advice on how to deal with risks and uncertainty, which vary in their degree of precision. Some examples are:

- "Dangers and unjustifiable risks to human health should be avoided." (Bundesregierung 2002: 51)
- "Risks of life-damaging events from human activity must be kept at very low levels. Technologies that threaten to cause serious and long-lasting damage to ecosystems or human health, at whatever level of risk[98], should be foregone." (Ekins et al. 2003: 176)
- "Technical risks with potentially catastrophic impacts on humans and the environment are to be avoided." (Kopfmüller et al. 2001: 236)[99].

All these are based, more or less, on the *precautionary principle*, being itself an important answer to conditions of risk and uncertainty within the sustainability context (e.g. UNCED 1992 b, WBGU 1998: 29, Council of the European Union 2006: 5). Instead of a single, established definition, there are many variants of the principle (EC 2000). In a rather weak formulation it holds that in view of "threats of serious or irreversible damage, lack of full scientific certainty shall not be used as a reason for postponing cost-effective measures to prevent environmental degradation" (UNCED 1992 b, Principle 15). Other, stronger versions imply that activities should be foregone if they are suspected of causing significant potential damage to human health or the environment. Most environmentalists tend to endorse a risk-averse view and favour a rather high degree of precaution.

Due to its stronger formulations the precautionary principle has come under attack[100] from some political theorists, for instance Cass Sunstein (2005). Sunstein criticises the principle for being incoherent (in its strongest forms), because there are risks on all sides of decision-making —not only the risks emanating from, say, a certain technology, but also in terms of the risks that come up in foregoing it (cf. also Wynne 1992: 124, Renn 2008: 79). Societal risks are not taken for their own sake, but because some benefit is expected from the activity associated with the risk (Renn 2008: 5). According to Sunstein, the precautionary principle

98 Note that the authors seem to conflate risk and probability here.

99 My translation.

100 For a summary of the debate, see Renn (2008: 81).

therefore "forbids the very steps that it requires" (2005: 4)[101]. On the other hand, advocates of the principle argue that it does not necessarily lead to banning a certain technology, for example, but rather calls for a gradual introduction until more knowledge is available. Its primary application field should therefore be situations of uncertainty and not of calculable risks (Renn 2008: 81, 91). Even Sunstein suggests the application of a restricted version of the precautionary principle ("anti-catastrophe principle") for cases where probabilities cannot be assigned to potentially catastrophic risks (2005: 5; resembling the principle proposed by Kopfmüller et al., see above; cf. also Birnbacher 1988: 155).

Research on risks within social science has found that the risk perception of laypeople is not necessarily in line with the "objective" calculation of risks as the product of harm and probability. Laypeople tend to be averse to particular risks that happen to call their attention, and unaware of others. Among the former there are, for instance, risks whose associated harm is particularly terrible, risks for which it is easy to think of cases where harm has materialised, and risks that are not taken voluntarily (Sunstein 2005: 5, Renn 2008: 16 ff., 93 ff., Heinrichs 2013 a: 229). Basing decision-making on "irrational" public demands invoking the precautionary principle would therefore potentially lead to excessive measures against certain risks and uncertainties, while foregoing opportunities and creating harm in other realms.

There is thus substantial controversy about how decision-making processes should work under conditions of risk and uncertainty (cf. Shrader-Frechette 1991, 1998, Sunstein 2005, Renn 2008: 284 ff., Aven/Renn 2010: 27, Pellizzoni 2010). Some suggest more reliance on technical expert judgements and cost-benefit analysis, while others highlight the political character of decisions involving risk and uncertainty because of the impossibility of being really objective in selecting risks (Renn 2008: 2 f.) and call for (participatory or deliberative) democratic procedures. Decision-making in cases of risk and uncertainty poses yet different problems when the distribution of risks and benefits is such that the benefits lie on the side of those taking a decision and the risks fall on others, for instance on future generations or remote places of the planet. Imposing significant risks on others without their consent, even if they

101 Even if Sunstein is in principle right here, I think he overdoes his point. It is easy to think of examples where there are no risks involved on some sides, at least if one does not conflate risks and opportunity costs. If I am, for instance, fearful of flying and stay at home instead of going on holiday with my friends, I may take into account high opportunity costs but not concrete risks. This is, of course, not the case if I decide to go by car instead of flying, which involves the (higher) risk of a car accident.

should also share the benefits, is morally highly problematic (Birnbacher 1988: 151, Shrader-Frechette 1998: 332, Caney 2010: 170; cf. also the discussion on the harm principle in section 4.2.1).
I will come back to the issue of risk, uncertainty and precaution in relation to environmental issues and consumption patterns in sections 4.2 and 4.3.5.

2.3.5 Sustainability, development and economic growth

The definition given in the Brundlandt Report cited before refers to "sustainable development", not to "sustainability". These two terms are often used as synonyms, yet it is worth having a closer look at the differences. While one of them is a process, the other is a status that is seen as desirable by its advocates. Moreover, "sustainable" means maintaining something and includes a notion of stability, whereas "development" means change, dynamic, and may also entail growth (Kopfmüller et al 2001: 32, Pearce 1988: 598). Thus, there is an inherent tension within the term sustainable development, which the term sustainability avoids from a linguistic but not necessarily from the conceptual perspective.[102]

Development

What is more, the term "development" is controversial. Some interpret it as primarily economic development, which is just another word for economic growth, which is again the growth of gross domestic product (GDP). This interpretation reflects the intention of the Brundtland Report to reconcile economic growth and environmental protection. The report uses "sustainable development" and "sustainable growth" nearly as synonyms (Petschow 2007). This is also the case with regard to the *Agenda 21*, which also contains the notion of "sustainable growth" (cf. UNCED 1992 a: chapter 4). Some theorists therefore consider the term sustainable development to be an oxymoron because, according to them, development involving economic growth cannot be sustainable due to the adverse envi-

102 Parris and Kates (2003: 561) try to resolve this ambiguity by listing different items that are *either* to be sustained (nature, life support, community) *or* to be developed (the economy, people, society).

ronmental impacts produced by economic activities (cf. Muraca 2009: 242). A plausible objection in defence of the term sustainable development has been raised by Daly: "When something grows it gets bigger. When something develops it gets different", so "development means qualitative improvement and growth means quantitative increase" (Daly 1996: 9; cf. also Georgescu-Roegen 1975: 363 f., Goodland/Daly 1996: 1002). However, in everyday language development often implies growth: for instance, the development of a child includes physical growth, besides many other things.

In an attempt to pacify this controversy, Kopfmüller et al. state that it is precisely the challenge raised by the idea of sustainable development—creating links between both its elements "sustainability" and "development" and realising both of them—which entails enormous innovative and creative potential (Kopfmüller et al. 2001: 32). Even if the Brundlandt Commission's interpretation of development mainly as economic growth may be challenged, it has to be acknowledged that the integration of environmental and development issues is indeed a core concern of the report.[103]

Economic growth

If one agrees that sustainability and development do not exclude each other, however, there still remains the question of whether sustainability and economic growth do[104]. The decades-old debate about economic growth has attracted new attention in recent years due to deteriorating forecasts with regard to climate change and in particular since the start of the financial crisis in 2008. At the same time, several recent visions of sustainability or sustainable development contain economic growth as one of their most important goals (e.g. UN 2015 b, Sachs 2015), and the EU links its

103 For instance in the following (WCED 1987: I/1/48): "The concept of sustainable development provides a framework for the integration of environment policies and development strategies—the term 'development' being used here in its broadest sense. The word is often taken to refer to the processes of economic and social change in the Third World. But the integration of environment and development is required in all countries, rich and poor."

104 For an overview of arguments in favour of the thesis that economic growth benefits the environment as well as arguments for the opposite thesis, see Neumayer (1999: 76 ff.).

sustainable development strategy to its "Europe 2020" strategy[105], which is a programme for "smart, sustainable, and inclusive growth" (EC 2010).

Economic growth is defined as the real growth of gross domestic product (GDP), which is the sum of the value of all final goods and services produced within an economy over a certain period of time. It was long seen as a good thing in general, providing a simple tool to assess whether the current development of an economy was positive. Reasons for its promotion are, among others: it is linked to other important parameters that directly influence the satisfaction of the population, in particular the unemployment rate (Spitznagel 2007: 69 f., Paqué 2010: 183 ff., Deutscher Bundestag 2013: 76 ff.); distributional issues can be solved more easily if what is to be distributed increases, so it is not necessary to take something away from some people if others are to get more (Kurz 2007: 21); some social systems, such as pension schemes, will have more difficulty managing the challenge of demographic change without economic growth (cf. Höpflinger 2010); and finally, it is hoped that continued growth can help to consolidate public households and reduce public debt (Paqué 2010: 204, Holzinger 2010: 33 ff.). Moreover, it is argued that growth is compulsory to a certain degree because of a monetary spiral of credits, money creation, investments and consumption, with the only alternative being economic de-growth (Binswanger 2006, 2009). However, there is significant controversy about whether these reasons to promote economic growth are still appropriate and the underlying mechanisms correct when economies become more mature, and whether the actual GDP growth rates in these economies are at all able to satisfy the expectations that are related to economic growth (cf. Rudolph 2007, Jackson 2009: 62 ff., 121 ff., Miegel 2010: 27 ff., Seidl/Zahrnt 2010).

The promotion of economic growth has therefore come under attack from several perspectives. On a conceptual level, it has been stated that infinite exponential economic growth is simply impossible in view of the numbers: an economy that grows at a rate of 4% per year would achieve the 1000-fold size of its original volume after 177 years. The possibility of such a development is widely seen as unlikely (Diefenbacher 2007: 33), but there are also contrary views (e.g. Neumayer 2000). Such calculations must be distinguished from similar claims related to the "size" of the economy in terms of its appropriation of natural resources. Environmental

105 See http://ec.europa.eu/environment/eussd/ (last accessed December 2015).

economists do not tire of insisting that the economy can only use a limited share of the natural resources that are provided by the Earth. The total amount of the Earth's resources places a physical limit on the global economy and human activities more generally (Meadows et al. 1972, Daly 1996: 33 ff., 48 ff., Costanza et al. 2012). These two aspects of economic development—the mere *number* describing it, expressed in terms of GDP change, and the development of the *physical extent* or "scale" of the economy within nature—should be kept separate carefully, which is often not the case within the debate on economic growth.

However, it has also been observed empirically that GDP growth is often accompanied by increasing resource use and pressures on the environment, in other words, an increased "scale" of the economy within nature. This is a major argument of those who criticise economic growth (cf. e.g. Hueting 1980, Bookchin 1989, Leipert 1991, Getzner 2009, Jackson 2009: 6 ff., Miegel 2010: 92 ff.). The database on the linkages between GDP and environmental impacts, however, shows diverging trends. There are cases where (certain types of) environmental degradation first become more relevant in growing economies and then reach a turning point and start decreasing after some time despite continued economic growth. This phenomenon, known as "decoupling", is the strategy of those who advocate "qualitative growth" (see below).

Further criticism about economic growth concerns social aspects. In cross-country analyses, researchers have found that there is no significant correlation between the level of wealth and happiness or well-being after a certain level of income has been reached (Frey/Stutzer 2002: 9 f., Layard 2005: 46). It therefore seems that at least beyond some level of wealth, well-being and economic growth start to decouple. While it should be noted that there is substantial controversy about the correctness of the interpretation of studies in this area[106], they have nonetheless spurred a debate about whether wealthy states should continue to pursue economic growth at all, as this might not translate into further increases in well-being. Moreover, several adverse social developments are often associated with economic growth, such as certain working conditions or the trend towards

106 For a very harsh critique, see an article featured by the Mises Institute which expresses the concern that governments might abuse the research results to build up "socialism" based on the alleged relationship between happiness and more egalitarian income levels (DiLorenzo 2011).

commuting, which lead to stress and have impacts on health (Leipert 1991, Miegel 2010).

In addition, GDP itself and preceding measures have been under attack for decades now: its interpretation as a major (or even the sole) indicator of welfare and its general informative value is being questioned for a number of reasons. First, GDP does not value the goods and services it counts. It also increases when services are demanded that indicate that something bad has happened, such as a car accident that incurs hospital costs and requires repair of the car, which obviously does not increase wealth. Second, it does not give information on the distribution of wealth within a country (Nussbaum 2011: 49). Third, it does not contain the costs of environmental degradation or the contributions of nature to the economy. And fourth, non-market activities or the "informal" part of the economy, particularly housework, are not considered. Further points can be added (see e.g. Diefenbacher and Habicht-Erenler 1991: 12 f.).[107]

Various economists, mostly within expert commissions set up by international organisations or countries (e.g. in France and Germany), have therefore presented alternative measures to GDP[108]. Proposals include sets of indicators, combined indicators of multiple components, and monetary indicators where, among other things, the benefits of housework and the costs of environmental degradation are included (cf. Deutscher Bundestag 2013: 302 ff.). Despite all the criticism, GDP growth continues to be one of the major political goals governments all over the world are committed to.

The scepticism related to economic growth is most widespread in industrialised, rich countries, which saw their highest growth rates decades ago. Most so-called developing countries today stress their right, despite the environmental problems involved, to catch up through economic growth in order to reduce poverty. It is therefore advisable, particularly for

107 For an—at least partial—defence of GDP, see for example Kassenboehmer and Schmidt (2011).

108 In France, the "Stiglitz-Sen-Fitoussi Commission" published its report in 2009 (see Stiglitz et al. 2009). In Germany, the Federal Environment Agency commissioned a study to compile a "National Welfare Index" (Diefenbacher et al. 2010), and a parliamentary commission developed a broad indicator set to measure welfare and progress (Deutscher Bundestag 2013). Examples besides these include the Index for Sustainable Economic Welfare (ISEW), the Genuine Progress Indicator (GPI), the Human Development Index (HDI), and the Happy Planet Index (HPI).

scholars and activists from wealthy countries, to be careful not to condemn economic growth in general.

The conflict about growth between "northern" environmentalists, moved by the discussion on the limits to growth, and "southern" efforts against poverty was already a major issue in the early sustainability debate in the 1970 s and 80 s (Dryzek 2005: 153, Michelsen/Adomßent 2014: 8 f.). Reconciling economic growth in order to reduce poverty with a reduction in environmental burdens was one of the major objectives of the Brundtland Report, which emphasised that poverty itself is an important contributor to the overexploitation of local environments[109]. The Brundtland report did even suggest that further economic growth in wealthy countries might be helpful in fulfilling that aim, arguing that as developing countries "are part of an interdependent world economy; their prospects also depend on the levels and patterns of growth in industrialized nations". Therefore, as claimed in the report, growth rates of 3-4 per cent could also be environmentally sustainable for industrialised countries "if industrialized nations can continue the recent shifts in the content of their growth towards less material- and energy-intensive activities and the improvement of their efficiency in using materials and energy" (WCED 1987: I/ 2/32; cf. also UNCED 1992 a: Nr. 33.3, 33.17). As stated above, economic growth continues to be part of many prominent conceptions of sustainability and, for instance, is part of the Sustainable Development Goals (UN 2015 b).

The different positions on the compatibility of sustainability and economic growth are outlined in the following. Because of the inadequacy of equating "developing" and industrialised countries in this regard, I restrict my analysis to the latter, which is also in line with the general limitation of this study to wealthy countries (cf. section 1).

---

109 See, for instance, the following statement from the report (WCED 1987: I/2/29): "As indicated earlier, development that is sustainable has to address the problem of the large number of people who live in absolute poverty—that is, who are unable to satisfy even the most basic of their needs. Poverty reduces people's capacity to use resources in a sustainable manner; it intensifies pressure on the environment. Most such absolute poverty is in developing countries; in many, it has been aggravated by the economic stagnation of the 1980 s. A necessary but not a sufficient condition for the elimination of absolute poverty is a relatively rapid rise in per capita incomes in the Third World. It is therefore essential that the stagnant or declining growth trends of this decade be reversed."

Compatibility of sustainability and different paths of economic development

Both an oversimplistic "limits to growth" position, which does not clearly differentiate between economic growth and the scale of the economy in the environment, and a position that idealises economic growth as a solution to all problems do not seem appropriate. In principle, these two positions still exist, but the debate is more differentiated now and based on more accurate empirical data. Also, there are not only two paths, growth or stagnation, but a diversity of positions are held by participants in the debate. Five different paths of proposed economic development are viable in this context:

- Business as usual: growth regardless of its "quality",
- "qualitative" growth without increasing or even decreasing negative impacts on the environment,
- zero growth, often named "steady state" or "stationary state",
- "de-growth", and
- ceasing to focus on economic growth as a relevant indicator.

While the first option is clearly not sustainable, as it disregards both the environmental dimension and social criticism of the existing pressures to grow, the remaining four can, in principle, all be compatible with sustainability. Whether they turn out to be sustainable in practice depends on their respective related empirical consequences.

Qualitative economic growth, which means GDP growth while an increase in negative impacts on the environment is avoided through their "decoupling"[110] from GDP, is the favoured strategy envisaged in many official documents on sustainability (e.g. WCED 1987: I/2/32-37, Bundesregierung 2002, Council of the European Union 2006)[111]. Lately, it has also been embraced in "Green Growth" or "Green Economy" strategies to

110 "Absolute" and "relative" decoupling are often distinguished between, with only absolute decoupling meaning an absolute reduction in adverse environmental impacts, while relative decoupling refers to a situation in which the environmental impacts are still growing, though at a slower rate than GDP (Jackson 2009: 67, EEA 2013 a: 39, Deutscher Bundestag 2013: 430). Where the term decoupling is used in this text, I mean absolute decoupling.

111 Note that the recently published Sustainable Development Goals are less bold on the issue of decoupling in the promotion of "sustained, inclusive and sustainable economic growth": a sub-goal cautiously asks to "[i]mprove progressively, through 2030, global resource efficiency in consumption and production and en-

overcome the economic crisis since 2008 onwards (e.g. OECD 2011 b, UNEP 2011 b; cf. section 3.1.5). Improving the "quality" of economic growth means making it less material and energy intensive by increasing efficiency and dematerialising economic activities. This way, it would not be necessary to abandon economic growth altogether; however, it is often stated that qualitative growth can only be moderate growth in order to allow technical progress to reduce environmental impacts despite increasing economic activity (Binswanger 2009: 12, Jänicke 2011: 15).

Whether decoupling strategies can be successful in all areas of environmental concern and on the global level is controversial. In some types of environmental pollution, such as sulphur dioxide emissions, industrialised countries have shown significant improvement in the level of pollution despite increases in GDP over the years (cf. chapter 1). For these environmental parameters, empirical data shows that after a certain period of increase in pollution, the trend is reversed and environmental pressures start to decrease with growing GDP. The resulting inverted U-shaped curve is known as the "Environmental Kuznets Curve" (EKC)[112]. This suggests that economic growth could also be part of the solution to environmental problems, rather than their cause, creating wealth and knowledge that can be devoted to environmental technologies (cf. the discussion in Neumayer 1999: 82 ff.).

However, an EKC has only been detected for some environmental quality indicators, in particular those which have a direct and local effect on health: mainly air emissions such as sulphur dioxide, particulate matter, nitrous oxides etc. (Dinda 2004: 441). For other environmental parameters no EKCs have been observed, for instance for resource use and greenhouse gas emissions (Jackson 2009: 71 ff., Steinberger et al. 2013)[113].

---

deavour to decouple economic growth from environmental degradation" (UN 2015 b).

112 Named after Simon Kuznets, who had claimed the discovery of a similar relationship between income inequality and economic growth in 1955.

113 Which does of course, from a logical point of view, not necessarily have to be so in the future. A study commissioned by the German Federal Environment Ministry presented models showing the possibility of reducing EU greenhouse gas emissions by 30% by 2020 while increasing economic growth (Jaeger et al. 2011). Whether all the technological measures mentioned in the report are desirable and will be effective is, however, an open question, as is whether these estimates are realistic. Jackson (2009: 81) shows that at a global economic growth rate of only 2% per year, to achieve the target of a maximum of 450 ppm carbon

Moreover, it is suspected that the "export" of polluting industries to other countries due to higher environmental standards in wealthy countries plays a major role as an explanatory factor for the existing cases of EKCs (Dinda 2004: 436 f., 449). This means that environmental degradation does not "disappear" with the growth of an economy beyond a certain point, but may at least partially be shifted to other economies at a lower stage of economic development. Thus, the feasibility of a *global* decrease in at least some environmental impacts at overall positive rates of economic growth may be questioned.

As an alternative, environmental economists have developed the concept of a *steady-state (or stationary-state) economy*, which does not grow further beyond a certain point, but remains the same in terms of volume[114]. Much earlier, classical economists already expected economic growth to come to a halt at some point in time, and most of them thought of ways to retard this—in their eyes—undesirable, but unavoidable situation (cf. Luks 1999: 179 ff.). A notable exception is the interpretation of the stationary state by John Stuart Mill, who also believed that both capital and population growth would cease unavoidably at some time, but called into question the use of infinite economic and industrial "progress". Mill did not see the stationary state as implying a halt to "human improvement", and regarded it as "a very considerable improvement on our present condition" (Mill 1909 [1848], book IV, chapter VI). Only with the rise of neoclassic economic theory did the idea of infinite economic growth become the standard it continues to be today, and the possibility of a stationary state was seen as something abnormal that needs to be avoided (Luks 2001: 152).

This standard view was confronted with the positive image of an "equilibrium state" in the 1970s in the context of the debate on the limits to growth (cf. Meadows et al. 1972: 170 ff.) and with the vision of a steady-state economy, most prominently advocated by Herman E. Daly (e.g. 1977, 1991, 1993, 1996). What appears to be problematic about these vi-

dioxide in the atmosphere by 2050, the carbon intensity (the CO_2 emissions per unit of GDP) would have to decrease to 1/21 of the global average in 2007, which is still 1/7 of the carbon intensity of Japan in that year.

114 Steady state in this sense is not to be confused with the term "steady state" as used commonly in neoclassical economics, where it refers to a situation characterised by a *constant rate of growth* of the relevant variables over time (Luks 2001: 175 ff.).

sions is in what sense the economy is supposed to come to a steady state: in terms of GDP development or use of natural resources? Daly is quite explicit in some instances that he is referring to natural resources, that is, the *physical* scale of the economy, as the following quotation shows (Daly 1996: 31 f.):

> *"Growth, as here used* [sic], *refers to an increase in the physical scale of the matter/energy throughput that sustains the economic activities of production and consumption of commodities. In an SSE* [steady-state economy] *the aggregate throughput is constant [...]. The other crucial feature in the definition of an SSE is that the constant level of throughput must be ecologically sustainable for a long future for a population living at a standard or per capita resource use that is sufficient for a good life. Note that an SSE is not defined in terms of gross national product. It is not to be thought of as 'zero growth in GNP'."*

However, he also does little to avoid confusion with economic development as conventionally measured by GDP (or GNP), for instance in his characterisation of GNP: "GNP is an index of throughput, not welfare. [...] To design national policies to maximize GNP is just not smart. It is practically equivalent to maximizing depletion and pollution." (Daly 1996: 41). Daly also states that "[t]he concept of a SSE [steady-state economy] is independent of GNP, and what happens to GNP in the SSE simply does not matter. The best thing to do with GNP is to forget it" (Daly 1993: 330). If GNP was an index of throughput in Daly's sense, it would be a measure for what characterises the steady-state economy he advocates, although he objects to this direct link. In consequence, a steady state of the physical extent of the economy would go along with a steady state of GNP (or GDP) (Luks 1999: 358 ff., see also Luks 2001: 37 f.).

After all, what Daly proposes focuses on the physical scale of the economy within nature, and the effect on economic (GDP) growth is merely shrugged off as an irrelevant consequence. It thus is questionable whether there are any current advocates of a steady state for GDP that is independent of other indicators at all.

A more radical perspective that has become popular in recent years among environmental and anti-globalisation activists is represented by the *"de-growth"* movement. This movement is very diverse and so are the theoretical approaches proposed in this context. De-growth, as the term seems to imply, could be something that just sounds better than "shrinkage", and refers to a decreasing GDP and/or decreasing throughput of environmental matter. However, as one of the most prominent thinkers of the movement makes clear,

> *"[t]he slogan of 'de-growth' is primarily designed to make it perfectly clear that we must abandon the goal of exponential growth, as that goal is promoted by nothing other than a quest for profits on the part of the owners of capital and has disastrous implications for the environment, and therefore for humanity. [...] De-growth is not, in my view, the same thing as negative growth. That expression is an absurd oxymoron, but it is a clear indication of the extent to which we are dominated by the imaginary of growth. [...] Strictly speaking, we should be talking at the theoretical level of 'a-growth', in the sense in which we speak of 'a-theism', rather than 'de-growth'."* (Latouche 2009: 7 f.)

In Latouche's view, de-growth is not an end in itself and does not necessarily relate to the growth or shrinkage of something, but is rather a label for a radical critique of the paradigm of economic growth (and of *development*, too), which aims to replace this paradigm with alternative models of society (Latouche 2009: 7, 9, cf. also Muraca 2009: 242, Martínez-Alier et al. 2010: 1742 f.).

However, there are also views within the de-growth debate that literally suggest a decrease in the economic throughput and of consumption and production, as the current scale of the economy within the environment is seen as too large and unsustainable (cf. Martínez-Alier et al. 2010). As de-growth is not an end in itself, it is however questionable whether such a decrease has a defined target at which this process would stop; in this sense, it is possible to combine the ideas of de-growth and steady state, which means that a process of de-growth of the economic scale of rich countries would after some time turn into a continuous state in which the economy has reached its "optimal" scale (cf. Martínez-Alier et al. 2010: 1744).

With regard to GDP, it seems that the intention of the de-growth idea is not primarily to achieve a decrease in this indicator (independently of others) but rather to ignore it. Even though it is likely that GDP shrinkage would result in a decrease in adverse environmental impacts, this cannot be guaranteed over the long term (van den Bergh 2011).

Both of the previously mentioned positions on the economic growth debate, steady state and de-growth, therefore seem to merge, at least with regard to GDP, with the last position presented here, which holds that those striving for sustainability should simply *stop paying much attention to GDP*. This position is most explicitly presented by van den Bergh (2011), labelled by him, in line with the above quotation by Latouche, as "a-growth":

> *"If one accepts that GDP (growth) is not a robust, reliable indicator of social welfare (progress) then the only solution is to ignore it and as a result be completely indifferent about GDP growth. GDP growth is good in some periods or for some countries, but unconditional growth is not a wise aim. GDP growth is not generally necessary or sufficient for progress. Neither is GDP degrowth necessary or sufficient for sustainability. Correlations between GDP and welfare or between GDP and environmental impact are not constant and fixed over time."*[115]

Van den Bergh sees the "dogmatic" fixation on GDP as a policy goal as an enormous information failure, which represents a significant hurdle to meaningful environmental policies, and should no longer be reproduced through media and education channels.

The same position—that GDP development is ultimately not a useful piece of information for those striving for sustainability—was expressed in the recent discussion on a "post-growth society". A sort of manifesto for such a society has been provided by Seidl and Zahrnt: in the sense defined by them, the "post-growth society" is characterised by not being dependent on economic growth, which ceases to be an important goal for politics, the economy and society. Thus, within this society, policies that aim to foster economic growth are abandoned; institutions that have so far depended on or are drivers for economic growth are redesigned in order to make them independent of growth; and the growth of energy and resource use is stopped and reduced to a sustainable level (Seidl/Zahrnt 2010: 34).

115 Cf. also Arrow et al. (1995: 521).

Synthesis

Table 3: Aims of different paths for economic development

	Aim with regard to GDP	**Aim with regard to environmental impacts / "scale" of the economy in nature**
Business as usual	growth	not focused on
Qualitative growth	moderate growth	constant or decreasing
Steady-state economy	not focused on, but anticipated that it will decrease or be constant at some level	constant at sustainable level
De-growth	not focused on, promotion of different models for society	decreasing
Stop paying attention to GDP	not focused on	decreasing until sustainable level is reached

Table 3 gives an overview of the different positions mentioned with regard to economic development that are held within the sustainability debate (and, with regard to the first one, outside it). To sum up, there remain only two positions in the sustainability debate with regard to economic (GDP) growth, as three of the positions outlined above converge into a single one: those in favour of "qualitative" growth advocate the further pursuit of economic growth and suggest that it can be decoupled from environmental impacts, or even claim that striving for more "green" technologies would bring further economic growth automatically; others, including those advocating a "steady state", "de-growth", or the "post-growth society", in principle all call for less attention on GDP as a relevant indicator and for focusing on the actual development of environmental impacts (and meaningful well-being indicators) instead.

While the idea of "qualitative" growth counts on the possibility of absolute decoupling of all relevant environmental impacts from GDP growth, which remains to be empirically proven, the latter position, which holds that GDP should not be an issue at all for sustainability, seems to be the more consistent one at least with regard to wealthy countries. After all,

GDP growth as such, independently of meaningful social and environmental indicators, does not play a role for the aims of sustainability. Positive and long-term impacts of GDP growth on socially relevant indicators such as the unemployment rate and the extent to which needs are met have been decreasing in significance in wealthy countries for quite some time. For poor countries, "qualitative" growth might be an interesting model, and even an increase in (certain) environmental impacts on these countries at the expense of rich ones would be justified, which is, however, not my focus here.

2.4 Interim summary

This chapter has provided a discussion on a number of relevant aspects of debate regarding the theory elements of liberalism, democracy, and sustainability. It has further outlined for each of the discussed aspects the position adopted within this study, which together form the basis of my analysis in the later chapters. The general strategy has been to adopt rather common positions that are compatible with mainstream theorising or constitute a minimal consensus for a whole range of theories, in order to build a maximally robust fundament for my argumentation in the subsequent chapters. This strategy has resulted in the following framework:

First, the framework consists in a liberal-democratic system. In its liberal component, this system is characterised by the equality of individuals' moral status; by individual liberty from state interference in one's own affairs, which is limited by the rights and liberties of others; by neutrality towards individual conceptions of the good; and should be understood as representing a political rather than an economic variant of liberalism. Further, in its democratic component, the system is understood as featuring a collective decision-making procedure that is based on equal voting rights and characterised by deliberation among the participants, which consists in an exchange of justifications they give each other to support their respective positions.

Taken together, the liberal and the democratic component build a complementary framework safeguarding individual rights and liberties as well as requiring those in power to respect the will of the majority. The demands of this framework constitute, on the one hand, a double boundary that political action taken in the name of sustainability needs to respect. On the other hand, it forms the basis of the demands of sustainability and can be a driver of these, as the protection of certain liberties and rights requires the introduction of environmental policies (cf. section 4.2), and as it is more likely to bring about a substantial level of environmental protection than other state systems.

Second, for the purposes of this study, sustainability is characterised by the following elements: a pragmatic and enlightened version of anthropocentrism; a conception of justice that focuses on meeting the needs of contemporaries and future generations rather than striving for distributive equality; the requirement of sustaining the "critical" stocks of natural capital (or, in other words, the requirement of not imposing excessive burdens on environmental carrying capacity) instead of postulating that the overall sum of natural capital or of all capital stocks, or the utility derived from these, is to be kept equal; and finally, by indifference regarding the question of economic growth as long as this does not jeopardise sustainability's aim of achieving the desirable state characterised by the previous elements. Sustainability in this sense is limited to a relatively "objective" minimum conception that focuses on preserving the integrity of the environment to safeguard meeting human needs now and in the future. At the same time, this does not deny the need for further-reaching, normative discussions on sustainability.

Thus far, little has been said with regard to the ways in which the aims of sustainability should be achieved. Different strategies are advocated within sustainability discourse: these mainly include solutions that focus on the development and broad penetration of better technologies on the one hand, and solutions that require changes in behaviour and lifestyles on the other, both of which are dealt with in the next chapter. Further, who is responsible and effectively able to implement these strategies is controversial—possible actors in the field include in particular individuals and public entities on various policy levels. This latter question will be dealt with in chapter 4.

3. Sufficiency and technical solutions

Introduction

In the early 1990 s, different strategies to achieve the aims of sustainability, particularly the environmental ones, emerged within sustainability discourse. Whereas in the English-language debate the strategies of "efficiency" and "sufficiency" are mainly distinguished between, the German debate features the additional term "consistency". The three strategies are sometimes described briefly with the words "better", "less", and "different" (BUND et al. 2008: 216, Sachs 2009).

While efficiency is a well-known approach aiming to optimise a cost–benefit ratio, the term "consistency" requires more explanation: It was coined and promoted by the sociologist Joseph Huber in the mid-nineties (cf. Huber 1995 a, 1995 b, 1999, 2000, 2011) and, in short, proposes developing newly adapted technologies that have little or no environmental impact, whereas efficiency in the present context aims at improving existing technologies. Both are industry-focused strategies that require only limited action by the end user, whereas sufficiency focuses on consumption patterns and behaviour-related aspects. Hence, as opposed to sufficiency, efficiency and consistency can be characterised as strategies of technological innovation (cf. Stengel 2011: 129). While these technology-focused strategies are implemented in many environmental policies in industrialised countries (for some examples, see section 3.1.3), the approach of sufficiency can be found less frequently in environmental policies and if so, it is rather a by-product.

Some authors have amended the troika of strategies with additional ones or have named them differently. For instance, ecological "resilience" (Ott/Döring 2004) or "de-commoditisation"[116] (Boulanger 2010) have been proposed instead of consistency, and "economic efficiency" (Müller 2009) or "regulatory strategies" (Stengel 2011: 129) have been suggested

116 Accordingly, this strategy aims at "acting one level up on the institutional context of consumption [...] [and] at decreasing the influence of commodities and, more generally, of the market institution in the way in which people satisfy their needs and desires." (Boulanger 2010: 6).

as additional fourth strategies. However, all these do not seem to improve on or fit into the existing categorisation—regulatory strategies, for example, rather run across the other ones and can be applied to all of them. It makes most sense for my purposes here to distinguish between primarily technical solutions and sufficiency. Still, even for these two, it turns out to be very difficult to define them and clearly separate them from each other. Most definition efforts hitherto have resulted in rather vague ideas, and a clear demarcation has not been found (cf. section 3.2.2). It has therefore been argued that instead of continuing the dispute about definitions of the terms, the discussion should return to its common aim of finding ways to achieve environmentally sound structural change (Diekmann 1999).

For the purpose of this study, however, it is necessary to develop at least a rough understanding of what is meant by sufficiency as opposed to technical solutions. A "hard" definition, which clearly allows us to distinguish cases of sufficiency from other cases, is not required, since it is not the intention here to play off one strategy against the other. Some participants in the debate think that the strategies are, in general, rivals, as there is a conflict of interest between them (Huber 1999: 12 f). Most, however, do think that the technology and the sufficiency paths are complements and need to be pursued in parallel because neglecting either of them would compromise achieving sustainability (e.g. BUND et al. 2008: 216 ff., Linz 2002: 10 ff.).

In the following, the discussions about the technological and the sufficiency strategies within sustainability discourse are presented, the meaning of the terms is discussed, and the limitations of each strategy are analysed. An interpretation of the term as used in this study is derived from a comprehensive literature overview on sufficiency. Finally, both the technology and the sufficiency paths are set in relation to similar or otherwise related concepts and ideas mostly from the sustainability context.

3.1 Technology and technical solutions

3.1.1 Technology in environmental discourse

The role of technology in relation to environmental problems is very controversial among environmentalists. On the one hand, technology is characterised as one of the causes of proliferating environmental change and adverse impacts, and is associated with potentially catastrophic environmental risks (e.g. nuclear energy). Technical progress has a long tradition in making it more and more effective and efficient in extracting natural resources, decreasing the price of these resources and increasing demand, while often leaving the surrounding ecosystems devastated. On the other hand, technical progress is hailed as a solution to environmental or, more broadly, sustainability problems (Kemp 1997: 1, Grunwald 2006: 39 ff., Beemsterboer/Kemp 2016: 71), as it has the potential of providing cleaner, more adapted technical solutions which produce less waste and emissions, while at the same time promising a higher quality of life by freeing humans from many diseases and toilsome tasks. Often one and the same technical solution, especially large-scale "techno-fixes" such as genetic or geo-engineering, is associated with both promises to solve an environmental issue once and for all as well as with enormous environmental risks. This double-edged role of technology can also be found with regard to consumption: technology can influence individuals' consumption by amplifying it or by promoting environmentally sound behaviour, for instance through feedback systems, or in other ways (Midden et al. 2007).

It is therefore no surprise that the question "For or against technology?" (Unseld 1991) seems to divide environmentalists into two camps. In the early days of environmentalism especially, those sceptical or even hostile towards technology represented a majority, while over the past 20 years technological optimists have gained ground (Mol/Jänicke 2009: 20).

A publication in the tradition of the first group is, for example, the famous *Limits to Growth* report (Meadows et al. 1972), which emphasised the inability of technical solutions to really *solve* certain problems and their tendency to postpone undesired effects only for a limited time instead:

> *"Over the past three hundred years, mankind has compiled an impressive record of pushing back the apparent limits to population and economic growth by a series of spectacular technological advances. Since the recent history of a large part of human society has been so continuously successful, it is quite natural that many people expect technological breakthroughs to go on raising physical ceilings indefinitely. These people speak about the future with resounding technological optimism. [...] Technology can relieve the symptoms of a problem without affecting the underlying causes. Faith in technology as the ultimate solution to all problems can thus divert our attention from the most fundamental problem—the problem of growth in a finite system —and prevent us from taking effective action to solve it."* (Meadows et al. 1972: 129, 154)

Another example is Ernst F. Schumacher's *Small is Beautiful*, which described modern technology as a force responsible for crises, unaware of limits, and inhumane, which produces problems faster than solutions. As an alternative to modern technology, Schumacher called for an "intermediate technology" that is affordable, less "violent" and less ecologically damaging, understandable, and can be repaired more easily (Schumacher 1973: 136 ff., 143 f., 166 ff.). As a final example, Hans Jonas' *Prinzip Verantwortung* (1979) should be mentioned, which focuses on the large-scale and long-term risks associated with technologies and which established Jonas' "Ecological Imperative"[117]. Further, the environmentalist criticism also interlinks with critical views on the social effects of technology. Such a view is expressed, for instance, in Ivan Illich's *Tools for Conviviality*[118] (1973), which is an overall critique of the industrial mode of production and the way in which it influences human relationships and abilities. Illich stated that the preferred way to solve problems created by science and technology had become the use of more scientific understanding and improved technology: "[T]he cure for polluted rivers is more costly nonpolluting detergents". In a rather dismal interpretation of technology, Illich concluded that instead of replacing slaves, "machines enslave men" (1973).

In the second, technology-friendly group, we find, for instance, environmental sociologist Joseph Huber (2000, 2008, 2011), designer William McDonough and chemist Michael Braungart, who are famous for their "Cradle to Cradle" concept (see below; McDonough 1998, McDonough/Braungart 1998, 2002), as well as the authors of *Factor Four* and *Natural Capitalism* (Weizsäcker et al. 1995, Hawken et al. 1999; cf. also White 2002). A quotation from an essay by philosopher Peter Sloterdijk, even though he is not an environmentalist or eager contributor to this debate, exemplifies the optimistic view of technology:

> *"Do we understand the conditions we find ourselves in correctly if we interpret the Planet and its biosphere as an unmultipliable One and proceed to grasp this as fixed limits that cannot be transgressed? [...] We do not yet know what development will become possible if the geosphere and biosphere*

117 "Act so that the effects of your action are compatible with the permanence of genuine human life on Earth." (Jonas 1979, commonly used translation).

118 Note that the German version of the book is entitled *Selbstbegrenzung: Eine politische Kritik der Technik*, which, directly translated into English, means: "Self-limitation: A political criticism of technology".

> *are advanced by an intelligent technosphere and the noosphere. It is not excluded a priori that this could spawn effects that would be equivalent to the Earth's multiplication. Technology has not yet said its last word. While to date we tend to view it from the angle of environmental destruction and bionegativity, this merely shows that in some respects it is still in its infancy. [...] By re-aligning the technosphere to meet homeotechnical and biomimetic standards in the course of time a completely different image of the interaction between the environment and technology would arise. We would find out what the Earth's body is capable of the moment humans switch in their dealing with Earth from exploitation over to coproduction. Down the path of mere exploitation, the Earth will forever remain a limited monad. Down the path of coproduction between nature and technology, it could become a hybrid planet on which more would be possible than conservative geologists believe."* (Sloterdijk 2010)

One of the publications mentioned representing the technology-friendly view, *Factor Four* by Ernst Ulrich von Weizsäcker and Amory and Hunter Lovins (1995), presents a more differentiated picture, however. While providing arguments for the potential of increased resource efficiency (see next section) to decrease environmental burdens, the authors also criticise overly optimistic, "cornucopian"[119] views on resources and technology, which were a counter-reaction to the *Limits to Growth* report, and point out that efficiency increases can only help to gain time (cf. Weizsäcker et al. 1995: 276 ff., 324). The same differentiated tone is present in the update *Factor Five* (Weizsäcker et al. 2009).

The suggested separation of environmentalists into those who are for or against technology thus seems to be outdated. Many adopt a more balanced view today, accepting the probem-solving potential of certain technical innovations, while still rejecting others, and at the same time emphasising that technical progress alone is not enough to deal with each and every environmental problem. The romantic image of the technology-free simple life is propagated less often, while certain developments that lead to a technological "arms race" are still criticised. The question therefore rather seems to be which technical developments should be scrutinised critically or even rejected by environmentalists, and where are the limitations of technology's problem-solving capacity? My focus here lies on

119 "Cornucopian" is often used as a label for technology optimists who deny environmental limits and/or that there are limits to the ability of science and technology to solve problems. Cornucopia (German: *Füllhorn*) means abundant natural supply—abundant resources and the unlimited capacity of the environment to absorb pollutants (cf. Dryzek 2005: 51).

technology's potential to provide solutions rather than the technologies that are viewed as inherently problematic.

Empirically, it can be observed that technical progress, in conjunction with supporting political innovations (Jänicke 2003, McNeill 2005: 373), has, in many respects, brought about enormous improvements to environmental quality in industrialised countries since the 1970 s. Air and water quality, which were at a low point in the second half of the 20th century in most European countries, have recovered since then, and a number of severe environmental problems such as acid rain and polluted rivers have become much less threatening.[120] Technical solutions to environmental problems themselves have also changed in character over time. From initial additive, "end-of-pipe" solutions, such as filters, a development took place towards more integrated and preventive solutions (Mol/Spaargaren 2000: 21, Mol/Jänicke 2009: 21).

However, as the famous "low-hanging fruits" of environmental protection have been harvested in rich countries, it now is increasingly difficult and cost-intensive to further reduce the environmental impact of these economies. Carbon dioxide from combustion processes, for instance, cannot simply be filtered or replaced (only sequestered and stored, which involves high costs and risks). In modern industrial societies, other typical environmental problems arise, such as increasing waste production, a high demand for motorised mobility and electricity, increasing land use, and risks and uncertainties related to new materials and toxic substances (Jänicke 2009: 33).

Technical solutions are still often highlighted as the silver bullet with which to overcome environmental problems (e.g. Sachs 2015: 85 f., 215) as they are considered the most appealing and least painful option, allegedly allowing us to circumvent distribution conflicts (Radkau 2008: 376). However, it is unclear to what extent technical innovations can adequately deal with the remaining or new problems. Before discussing the limitations of technical solutions, the next sections will shed more light on the term "technology" itself and on the kind of technical solutions advocated in the environmental context.

---

120 It should, however, be noted that to some degree, environmental improvement in wealthy countries with high standards for air and water quality is also due to the exportation of pollution-intensive production processes to poorer countries (cf. the discussion on "Environmental Kuznets Curves" in section 2.3.5).

3.1.2 The term "technology"

According to the *Routledge Encyclopedia of Philosophy*, technology[121] is usually conceived as "the transformation or manipulation of nature (the existing physical (material) and biological environments) to satisfy human needs and goals" (Kroes 1998: 285). Technology in this sense can result in a technical artefact, which is likely to be the first image that comes to mind if someone thinks of the term. However, this interpretation does not encompass everything that is commonly subsumed under the term technology, and it is too broad at the same time (Kroes 1998: 285). The focus on artefacts, for instance, does not take into account the immaterial aspects of technology, such as the know-how underlying it, the incorporated software, or its organised management and control structures and processes (Balabanian 1980: 8, Kroes 1998: 285). On the other hand, if all artefacts are seen as technology, that is, any object that emerges from human intervention with nature, this is also problematic as it raises issues about the demarcation between the artificial and the natural, and about the relationship between humans and nature. The demarcation problem, as the *Routledge Encyclopedia* states, therefore remains an open issue (Kroes 1998: 285).

The distinction between something that belongs to technology and something that does not thus often seems to be based on intuitive decisions. This becomes obvious in the treatment of "technology" within environmental science: In the environmental context, technology is not used in the strictly formal sense of the word. Instead, the discussion focuses on hi-tech solutions distingushed only vaguely from rather basic technologies that have been in use for decades or even centuries. By means of an example, there are certainly few environmentalists who would oppose the use of bicycles. The bicycle is the archetype of what Schumacher has called

121 In German, in this context not only is "Technologie" often used, but so is the more encompassing term *"Technik"*. *Technik* is a closer relative of the Ancient Greek word *techne* (art, craft) and is similar to, but not the same as, the English terms "technics" and "technique". It has mainly two meanings: one relates, similarly to technology, to the use of technical artefacts ("hardware"), and the other emphasises the procedural aspect of *Technik* (Grunwald 2008: 41 ff.), which is expressed in the English word "technique". In order not to complicate things unnecessarily, I focus in this chapter on the term technology. Where I use the word "technical" (e.g. in "technical solutions"), I use it in the sense expressed by technology.

"intermediate technology" (see above)—technology that is rather simple, decentral, easily repairable and suitable for maintenance, economical in its use of scarce resources, relatively cheap and "designed to serve the human person instead of making him the servant of machines" (Schumacher 1973: 143, 166 ff.). More recent visions of environmentally sound technology, however, suggest much more advanced kinds of technical solutions, precisely because there are doubts that simple technologies are always capable of using resources economically. In the following, I will therefore try to describe the kind of "technical" solutions advocated in the environmental context.

3.1.3 Technical solutions: efficiency and environmentally adapted technologies

A technical solution, in the words of ecologist Garrett Hardin, is "one that requires a change only in the techniques of the natural sciences, demanding little or nothing in the way of change in human values or ideas of morality" (Hardin 1968: 1243). Hence, the technical approach primarily relates to engineers and the production side, while the end user is not or only little affected, apart from in terms of the potentially higher costs a new technology entails.

As mentioned above, technical solutions to sustainability problems in principle include the two approaches of making existing technologies or activities more eco-efficient and of inventing new, ecologically adapted technologies ("consistency"). The distinction is not always straightforward, as it is often difficult to draw a clear line between improvements to an existing technology and the point where it starts to be a new one. Therefore, and in particular because such a clear distinction is not relevant here, this study summarises both approaches under the headline of "technology". Both are briefly explained below though.

Efficiency, in general, refers to the optimisation of an input-output or, more precisely, of a cost-benefit ratio. It must not be confused with effectiveness, which focuses on the achievement of targets but does not take costs into account. As efficiency deals with a ratio, it is not a metric in absolute terms. Thus, something becomes more efficient when either costs are decreased with benefits remaining the same, when benefits are increased at equal costs, or when costs are decreased and benefits are increased at the same time. As eco-efficiency it refers, for example, to the reduction of harmful emissions of a production site, while the amount of manufactured products remains constant, or to the reduced energy consumption of a product, while its functionality is not impaired.

Princen (2005) gives an entertaining overview of the cultural history of efficiency and the triumphal ascent of the idea. He finds that "[h]aving become nearly synonymous with 'productive' or 'useful', even 'good', efficiency has spilled into nearly all realms of modern life. […] It has become a social goal in its own right, equating with all that is desirable […]" (Princen 2005: 50). Indeed, efficiency has many advocates among politicians, industry, scientists as well as civil society groups. Some authors suggest focusing all efforts on an "efficiency revolution", that is, an improvement in a nation's (or the world's) overall energy efficiency or resource productivity by a factor of four, five, ten, "X" or "Y".[122] In the European Union and other regions or countries, there are manifold political instruments in place that aim at improving the eco-efficiency of industrial plants or products. An example are minimum energy efficiency standards for energy-using products, implemented in the EU through the Ecodesign Directive and its daughter regulations. These regulations set specific energy efficiency requirements for selected products.

As mentioned, the idea of ecologically adapted technologies is usually referred to as "consistency" in the German-language sustainability debate. It focuses on the material quality of technologies and mass flows, and proposes that these should generally be "consistent" or compatible with ecological cycles of materials. Unlike efficiency, consistency thus does not aim at reduction but at integrating industrial material and energy flows into natural flows as far as possible, and at closing material cycles for those irreplaceable materials that are ecologically harmful (Stengel 2011: 131 f., Huber 2000: 280). Examples in the sense of this approach are renewable energies or the use of materials that can be recycled in a "closed-loop" without losing their quality. Huber, who coined the term, describes it as "a strategy of qualitative change of the industrial metabolism by modernizing the basic structures of technology and products, allowing for a permanent turnover of material flows on a large scale and in large volumes" (Huber 2000: 280).

---

122 Cf. Weizsäcker et al. (1995, 2009), Schmidt-Bleek (1998), http://www.faktor-x.info/, http://www.factory-magazin.de/, http://www.factor10-institute.org (all last accessed June 2012).

The idea hence embodies the dream of a fully environmentally sound industrial metabolism, in which "scale" does not play a role any longer since industry does not have any or has only little negative environmental impact.[123] The same idea is in principle expressed by the "Cradle to Cradle" concept proposed by McDonough and Braungart (1998, 2002):

> *"Consider the cherry tree. It makes thousands of blossoms just so that another tree might germinate, take root, and grow. Who would notice piles of cherry blossoms littering the ground in the spring and think, "How inefficient and wasteful"? The tree's abundance is useful and safe. After falling to the ground, the blossoms return to the soil and become nutrients for the surrounding environment. Every last particle contributes in some way to the health of a thriving ecosystem. "Waste equals food" —the first principle of the Next Industrial Revolution."* (McDonough/Braungart 1998)

More specifically, "Cradle to Cradle" suggests designing products in a way that considers their whole life cycle not only until they become waste ("cradle to grave"), but until they can be converted into new products, or go back safely into the ground as organic material.

Policies in this sense, even though far from establishing perfectly environmentally sound material cycles, exist worldwide, for instance in the form of recycling requirements. A recent further step in this direction is the EU Action Plan for a "Circular Economy" (EC 2015).

3.1.4 Limitations of technical solutions

To what extent technical solutions can really solve environmental problems is, as already mentioned, an intensely discussed issue among environmentalists. Sometimes, the claim that technologies alone will be able to solve "the environmental crisis" is propagated simply as such, as a claim or belief (e.g. Sachs 2015: 215). Sometimes, it is backed by technology assessments, calculations, and scenarios (e.g. Pacala/Socolow 2004), which do, however, all contain their assumptions and simplifications, and often ignore the undesirable side effects of the technologies advocated (see be-

---

123 To be sure, Huber reminds us on one occasion that "there still are limits to growth, and a strategy of consistency should not lead one to expect a boundless land of milk and honey any more than the other strategies [efficiency and sufficiency] do" (2000: 282). However, the expectations related to the strategy as raised by its proponents are still very high.

low). It is questionable whether the issue is not just too complex and the related uncertainties too great to allow for an answer to this question based on forecasts. For example, not only do technological advances play a role, but so does the development of available economic resources and of human desires. As Wynne (1992: 111) notes, a

> *"difficult [...] question is whether environmentally sustainable futures are feasible even if we assume the most efficient systems of production to be universally in place tomorrow. Might not growing consumption and production simply swallow up the advances provided by those imagined technical utopias?"*

Thus, Sloterdijk (2010) is actually right in saying that we "do not yet know what development will become possible" and that it "is not excluded a priori" that effects equal to the multiplication of the planet will be generated (cf. section 3.1.1). However, his implicit conclusion that we should wait for such a development instead of accepting some limits to growth is not at all convincing. According to Ekins, the extent to which technologies can resolve environmental problems cannot be determined theoretically, but this is "essentially an empirical question" (Ekins 1993: 280).

A number of limitations to the problem-solving potential of technologies in the environmental realm can be identified[124]. In the following, I will describe these and, as far as possible, back them up with empirical observations.

Scale

An issue which is quite obvious from a theoretical view, which according to Daly can be labelled as a "scale"[125] constraint, applies to both environmental sinks and sources. The available sinks place a limit on the application of even the most environmentally sound technologies, since they still have *some* adverse impact (see below) which cannot be imposed on ecosystems in an unlimited way. Even many "natural" processes would harm ecosystems if applied to a limited system in large quantities. For instance, a lake will collapse under certain conditions if nutrients—for in-

---

124 Cf. similar discussions in Schor (2005 a: 310), O'Neill et al. (2010: 26 ff.), Schneidewind/Palzkill-Vorbeck (2011: 8 ff.), Stengel (2011: 131 ff.), EEA (2012 b: 7 f.), Ekardt (2016).

125 On the idea of "scale", cf. also section 2.3.5.

stance, from McDonough's and Braungart's cherry tree—are supplied in excess after all buffer capacities have been used up[126]. As Paracelsus is supposed to have said with regard to human toxicology, "nothing is without poison and only the dose permits something not to be poisonous."

Also, there is the constraint of available—renewable and non-renewable—resources. All technologies require resources such as raw materials and suitable space, but appropriate sites, for instance, for the installation of renewable energy technologies are limited (Sachs 2009: 39). Hydropower and wind turbines depend on certain conditions and cannot be installed randomly on any spot of ground that is free. In Central Europe finding suitable spots for hydropower and increasingly also for wind power is becoming more and more difficult, and conflicts of target with nature protection, aesthetic concerns and local resident interests are common. Bio-based fuels require vast cropland areas and generally compete with food production.

Thus, technical strategies have to acknowledge some physical limits related to the scale of the planet (Schor 2005 a: 310, O'Neill et al. 2010: 28, Stengel 2011: 139). Technical optimists would probably say in response to this that if one limit is reached, the market would come up with something else. However, again with regard to the example of renewable energy, the scarcities of required land and resources are already perceivable, and a completely new approach that will save us from these scarcities (at least for some time) is not visible.

No sustainable and no 100% environmentally sound technologies

An article by environmental innovation researcher René Kemp is titled *Sustainable Technologies do not exist!* (Kemp 2010). What Kemp refers to is that sustainability is a systemic matter and particular technologies or organisational approaches cannot be labelled as sustainable, as goods are

---

126 Also cf. the interesting comments by Jensen et al. (2011: 686 f.) on "waste equals food" thinking, challenging the view that ecosystems produce little or no waste and that this is always a good thing: "[…] is internal or externally deposited waste actually a bad thing, or is it simply a human construct that has no meaning in ecology? In some cases, 'waste' is arguably good. […] Waste is only seen as bad when it impinges on or threatens the long-term existence of a given organism."

produced by chains of practices and their impact also depends on the way they are used, electricity is generated and waste streams are managed at the end of their life cycles. The assessment of the sustainability of a good, technology or practice thus has to take into account use and its contexts (Kemp 2010: 8, Beemsterboer/Kemp 2016: 72).[127] In addition, Kemp notes that each technology or production and consumption activity has inherent environmental impacts, and they can also have adverse social effects: "Only a small proportion of the innovations would satisfy the criterion of it being Pareto superior, which means that no one is worse off. All radical innovations would not satisfy this criterion." (Kemp 2010: 8; cf. also Kemp 1997: 12). Fourty years ago, the authors of *Limits to Growth* called this the "side effects" of new technologies (Meadows et al. 1972: 146 ff.).

An obvious example is, again, that of renewable energies. Each of the renewable energy technologies developed so far has new environmental and/or social impacts. Besides the problem of limited space mentioned above, lately the promotion of biofuels especially has had strong adverse effects on world markets of some crops that are essential nutrients in poorer regions. In addition, pesticide use and other ecologically harmful operations during the cultivation of energy crops are an environmental issue. Such side effects also decrease the public acceptance of technical solutions. Another example is the idea of fully closing material cycles: Even if good practices exist today, such as biodegradable textiles and plastics, these also require energy and raw materials in their production, and therefore simply produce other side effects, even if they are minor and not as bad. New technologies just displacing adverse effects from one environmental medium to another should be avoided (BUND et al. 2008: 106). Besides that, it is far from clear whether such best practice examples can be scaled up and cover a relevant share of the market without decreasing the overall market volume at the same time (see previous point).

In order to provide empirical evidence of the technological feasibility of the "consistency" path, Huber undertook a survey on technological environmental innovations in order to demonstrate empirically that new technologies "tend to satisfy criteria of environmental sustainability ever

127 Cf., as opposed to this, the merely quantitative definition suggested by Weaver et al. (2000: 20): "Sustainable technologies would be ones capable of meeting 'needs' using only a fraction—less than a tenth and maybe only a fiftieth—of the 'eco-capacity' used by today's technologies."

more" (Huber 2008). However, among the conclusions nothing is said about the extent to which this is really the case or about the feasibility of truly "consistent" technologies in the future, and the examples given in the study (e.g. carbon capture and storage, renewable energies, transgenic biochemistry, substitution of hazardous chemicals with more benign low-impact substances, ultra-light and ultra-strong new materials) all involve risks and side effects.

Together with the aforementioned "scale" constraint, it therefore becomes clear that the dream of "consistency" and "Cradle to Cradle" advocates of establishing fully ecologically adapted technologies that allow us to disregard the turnover or "throughput" of material flows therefore remains a theoretical construct.

Rebound effect

The most well-known of the limitations discussed here is the "rebound effect". It describes a phenomenon already discovered by economist William Stanley Jevons in the 19th century, which is therefore also called the "Jevons' Paradox". Jevons found out that the more efficient use of coal led to an absolute increase in demand instead of lowering it as the use of coal became more profitable (Alcott 2005, Luks 2001: 148). Sometimes such over-compensation of efficiency gains, as opposed to merely partial or full compensation, is also called the "backfire effect".

In a comprehensive study reviewing the literature on rebound effects, Jenkins et al. (2011) distinguish micro and macroeconomic rebound mechanisms. The former are further split up into direct and indirect effects. Rebound effects have been analysed almost exclusively with regard to energy use, thus the following examples focus on energy.[128]

Direct microeconomic rebound effects are related to the application where an efficiency improvement occurs itself. Higher efficiency decreases the (implicit) cost of the application and can therefore give rise to a higher demand from a single consumer, household or firm. Examples in the field of energy efficiency are bigger energy-using devices such as fridges or TVs, a higher number of devices, extended use per day such as a longer distance travelled by car, or an increase in functions that raises

128 Maxwell et al. (2011) mention a number of other examples such as paper use.

energy use. Where direct rebound effects are avoided, the efficiency gain results in lower costs for electricity or other resources. When the money saved is spent on other resource-intensive consumption activities, indirect rebound effects are the outcome.

Macroeconomic rebound effects are the result of price effects and the price elasticities of demand. When, for example, a considerable number of households uses less energy, the total energy demand drops and so does, ceteris paribus, the energy price. Lower prices lead to an increase in demand, so that the reduction in energy use by some members of society will be offset in part or fully by others using more. This mechanism exists analogically at global level, when whole nations succeed in reducing their energy demand. The distinction between direct, indirect and macroeconomic rebound effects shows that much of the debate on these effects is a debate about system boundaries (Dahmus 2014: 893)—what is ultimately important from an environmental perspective is the overall development of absolute energy and other resource use, and not the narrow focus on single appliances or processes.

Controversy exists about the quantitative relevance of these effects and the related implications for efficiency policies. Examples of backfire effects have led some commentators to the conclusion that "since greater efficiency, ceteris paribus and given latent demand, must raise, not lower, environmental impact, efficiency policies are wrong" (Alcott 2005). Others, in a less pessimistic way, have stated that success in rationalisation, that is, improving efficiency, helps to save time, but is ultimately not effective (Sachs 1993: 69). However, according to empirical analyses, most rebound effects are usually far from compensating or even overcompensating for efficiency gains. Direct rebound effects for households in industrialised countries are estimated to be in the order of 10-30% of energy efficiency savings (Jenkins et al. 2011: 14). Macroeconomic effects could, as studies suggest, be much more relevant in comparison to microeconomic effects, compensating for up to 40% of the efficiency gains so that the sum, including direct effects, is around 50% (Barker et al. 2009: 425; see also Jenkins et al. 2011). The magnitude of these figures is open to interpretation: in a recent article in *Nature*, Gillingham et al. (2013) state that the rebound effect is overplayed and has become a distraction, judging the combined effects to be "small"[129]. The major concern put forward in the article is that energy efficiency policies are perceived as ineffective and might therefore be taken as an excuse for abandoning them, which is vigorously rejected by the authors. Indeed, the rebound effect does not make efficiency measures obsolete; it rather indicates that they are not enough and calls for complementary measures that are not subject to rebound effects, such as carbon pricing and other price incentives (Gillingham et al. 2013: 476, Dahmus 2014: 893; cf. also Weizsäcker et al. 2009: 309 ff.).

"No technical solution problems"

As Hardin has pointed out in his famous work on *The Tragedy of the Commons* (1968), a class of problems exists that can be called "no technical solution problems". Hardin discusses in particular the problem of overpopulation in this context, concluding that "[a] finite world can support only a finite population; therefore, population growth must eventually equal zero" (Hardin 1968: 1243).

Other examples more directly related to environmental impacts are mentioned by other authors. Stengel notes that neither "consistency" nor efficiency approaches can deal with the ecologic impacts tied to a global drove of billions of cattle or the overfishing of the seas (Stengel 2011: 132). Boulanger, who refers to similar examples, adds that, quite on the contrary, in these domains technical innovation rather goes in the opposite direction towards more and more intensive methods of harvesting. Therefore, some resources can only be protected by restricting their extraction or final consumption (Boulanger 2010: 6).

Available time and the cost involved

Finally, concerns exist that technical solutions might not be able to solve the most pressing environmental problems within the available time. Fundamental innovations and structural change in the sense of the "consistency" point of view will require much more time to be realised and in particular to be mainstreamed (Linz 2002: 11, Voget 2009: 213, Stengel 2011:

129 Actually, Gillingham et al. (2013) refer to studies that indicate even higher numbers than those mentioned above, estimating combined effects to be up to 60%.

132). This point is completely disregarded by e.g. "Cradle to Cradle" advocates (cf. McDonough/Braungart 2002).

With regard to climate change, Pacala and Socolow already claimed in 2004 that humanity would be able to solve the climate problem within 50 years by applying solely existing technology (Pacala/Socolow 2004). However, they also include options such as reducing car size and power and reduced car use in their portfolio of technical approaches, which I would clearly classify as non-technical (i.e. sufficiency) options. Further, the options they considered also include highly controversial technologies, such as nuclear energy and carbon sequestration, which create other environmental risks, as well as increased use of biofuels (corresponding to one-sixth of the world's cropland or 50 times the production of 2004), which would generate massive conflicts of interest, as discussed above. Most of these technologies therefore do not qualify as those fully clean and adapted technologies industrial ecologists dream of. Furthermore, even the "imperfect" technologies that are available come at a certain cost, in particular if they involve significant changes to infrastructure such as electric mobility and renewable energy sources.

3.1.5 Theories and political programmes with a focus on technology

Apart from the rather narrow technical approaches of efficiency and environmentally adapted technologies, there are whole theories, research areas and political programmes which have traditionally focused on technical solutions to environmental problems. Most notably, there are to mention Ecological Modernisation Theory and the research field of Industrial Ecology; further, there are political programmes and ideas such as a new "(Green) Industrial Revolution", the "Green New Deal", and the "Green (or Blue) Economy". It is worth presenting them here briefly because it is mainly within these fields where technical solutions to environmental problems are discussed, analysed and propagated. Lately, however, researchers in these fields have also increasingly emphasised the role of consumption patterns and behaviour, as well as organisational approaches reaching beyond the narrow focus on technologies, as this section will show.

Ecological Modernisation

The idea of Ecological Modernisation dates back to the early 1980 s, when it was still conceived as a political programme "pressing for far-reaching environmental reform" (Mol/Jänicke 2009: 18). Only later did it become a theory of environmental sociology, which is built on the assumption that a long-term transformation process towards more environmentally sound production patterns has emerged in industrialised countries. The theory takes this transformation process and related institutions and policies as its object of analysis (Spaargaren/van Vliet 2000). Furthermore, it assumes that the change required to deal with environmental problems can be realised in principle with the existing institutions and within the existing economic logic, even though a certain degree of structural transformation is also advocated (Mol/Spaargaren 2000: 19, Barry/Doran 2006: 254, Langhelle 2000: 305 f.). It is not assumed the transformation will happen based on market forces alone, but an important role is assigned to coordinated policy intervention in cooperation with industry. A central message of Ecological Modernisation is nevertheless "that there is money in it for business" (Dryzek 2005: 167).

According to its proponents, Ecological Modernisation Theory emerged as a counter-perspective to the more radical demodernisation debates that dominated the environmental discourse among environmental movements and social scientists in the 1970 s (Mol/Spaargaren 2000: 19, Mol/Jänicke 2009: 20; cf. section 3.1.1). That is probably why the theory has long been perceived as a purely technocratic programme and has been criticised as overly optimistic towards the problem-solving capacity of technology. This criticism is today fiercely rejected by Ecological Modernisation researchers (cf. Mol/Spaargaren 2000: 25, Spaargaren 2000: 324, Mol/Jänicke 2009: 20 f.; cf. also Dryzek 2005: 168). Not only are there different strands within the debate—some have a clearer focus on technology and the production side (e.g. Huber 2000, 2008, 2011), while others focus, for example, more on political and institutional development (e.g. Jänicke 2009)—but there have also been efforts in recent years to integrate the debate on lifestyles and consumption patterns into the theory (see Spaargaren/van Vliet 2000, Spaargaren 2000, 2009, Spaargaren/Cohen 2009, Mol et al. 2009: 10). Still, it can be said that the bulk of Ecological Modernisation research, even if "sustainable consumption"[130] has started to play a role within it, "is explicit about not attempting to limit overall levels of consumption" (Barry/Doran 2006: 260).

Industrial Ecology

Industrial Ecology refers to the more natural science and engineering-based sister discourse of Ecological Modernisation Theory. It focuses more on applications, since a large part of the research being done within the field analyses concrete proposals of how to measure or decrease the environmental impacts of production methods and products by means of technological change. Some of the principles of Industrial Ecology, for example the approach of looking at entire product life cycles, are by now well-established in environmental policies. Many different, broad and narrow, conceptions of Industrial Ecology exist—it is seen, among other things, as a paradigm, a research discipline, a metaphor (e.g. Socolow 1994), a notion that is "close to the concept of consistency" (Huber 2000: 281), or "the ecology of industry" (Jensen et al. 2011)—, and researchers in the field keep themselves busy with the task of fathoming out its boundaries (e.g. Lifset 1998, Hertwich 2005, Wells/Zapata 2012).

Although the discourse was historically exclusively focused on technology and production, initiatives to include consumption and lifestyles into the perspective of Industrial Ecology were already launched ten years ago. In a special issue of the *Journal of Industrial Ecology* on sustainable consumption from 2005, the editor stated that a couple of years earlier dealing with consumption under the label of Industrial Ecology was still perceived as "daring" by the community (Hertwich 2005[131]). Indeed, the "industrial"

---

130 Cf. section 3.2.5.

131 See also the entire special issue of the journal. Note that Socolow (1994: 4) had already written ten years before: "The focus of attention has long been on production facilities—conveniently far from the point of consumer involvement. By contrast, industrial ecology emphasises the management of products throughout their useful life and beyond and calls attention to dispersed sources of pollution, such as agricultural chemicals, household wastes, and the chaff resulting from the expected degradation of products like outdoor paints, roofing materials, and brake linings. The importance of the consumer is unmistakable." This is, however, not at all reflected in the comprehensive *Handbook of Industrial Ecology*, for instance (Ayres/Ayres 2002).

in Industrial Ecology does not support the inclusion of social and behaviour-related aspects of consumption in the analytical focus of the field (cf. Lifset 1998: 2). Nevertheless, there are also many measurable and engineering-related aspects of consumption that are proper subjects of study within Industrial Ecology.

Political programmes with a focus on technology

Further, there are a number of ideas that have emerged recently within environmental discourse that focus on technical solutions, which tend to have the character of political programmes supposed to implement the ideas developed within Ecological Modernisation Theory and Industrial Ecology. These ideas have all appeared within a relatively short time span —sometimes they are even mentioned together within the same political strategy papers—and received advocacy in particular during the aftermath of the recent global financial crisis.

The most obviously technical one, the idea of a new, green "Industrial Revolution" (McDonough/Braungart 1998, Hawken et al. 1999, White 2002, BMU 2008, Jänicke/Jacob 2009, Rifkin 2011), is inspired by earlier industrial revolutions that were triggered by fundamental changes with regard to the main energy and material sources, such as the use of coal and steel and, later, the widespread use of electricity and oil. The agenda for the next Industrial Revolution is therefore to again realise such a fundamental shift in order to overcome our dependence on fossil fuels and decouple resource use from economic growth (BMU 2008: 6 f.). Besides the technical challenge involved, the notion of an "industrial revolution" also comprises a broader societal transformation, including necessary political steering efforts (Jänicke/Jacob 2009: 11 ff.). This, according to the German Ministry of the Environment, makes the next Industrial Revolution a "political programme" (BMU 2008: 7). The notion has, however, also been criticised as deceptive: If one believes that a new technical revolution can be expected soon, this decreases the felt urgency to act now with available means; the notion thus wrongly suggests that the required changes will happen within a sphere of industry and technology alone, without affecting consumption patterns (Diefenbacher 2008).

The idea of a "Green" or "Sustainable New Deal" (Heinrich-Böll-Stiftung 2009, Sustainable Development Commission 2009, UNEP 2009) goes back to the "New Deal" policy in the US during the Great Depression of the 1930 s, which consisted of a comprehensive series of measures to overcome the economic crisis and put people back to work. The term Green New Deal became popular in the context of the financial and economic crisis from 2008 onwards. Proponents of the idea argue that the current crisis is twofold, relating to not only the financial system but also the global environment. This should be seen as a chance to bring about fundamental change: Financial programmes that are set up by governments worldwide should aim at environmentally beneficial structural change in order to fuel a green industrial revolution (Heinrich-Böll-Stiftung 2009: 7). The Green New Deal has three objectives: It is supposed to make a major contribution to reviving the world economy and creating jobs; it is intended to "promote sustainable and inclusive growth" and the achievement of the Millennium Development Goals (in particular with regard to ending poverty); and, finally, it aims at reducing ecosystem degradation and the dependency on fossil fuels (UNEP 2009: 4). As the idea of a Green New Deal has, first of all, a clear focus on "reviving the economy", it can be seen as a technocentric programme, since decreasing consumption levels is not beneficial to the aim of increasing economic growth. Existing political strategy papers (e.g. Heinrich-Böll-Stiftung 2009, UNEP 2009), however, also suggest some policy instruments that can have an effect on overall consumption, such as removing "perverse subsidies" that lead to the overexploitation of e.g. fish stocks and designing tax and incentive systems in an environmentally responsible way.

The most recent concept highlighted in political strategy papers is the "Green Economy" (see e.g. UNEP 2011 b, EEA 2012 a). The United Nations Environment Programme defines a Green Economy as one that brings about "improved human well-being and social equity, while significantly reducing environmental risks and ecological scarcities" (UNEP 2011 b: 16). In this sense, the idea seems to be nothing more than another "buzzword" supposed to replace the concept of sustainability (Brand 2012; cf. chapter 2.3). If we look in more detail at the respective UNEP report (UNEP 2011 b), the Green Economy seems, like the Green New Deal, to be very much focused on investments, economic growth and bringing people into jobs. On the other hand, besides investment and industry-side measures, it also contains a broad range of policy proposals that aim at "investing in natural capital", for example by limiting the rate of exploitation of fish stocks, as well as some rather sufficiency-like proposals, such as promoting non-motorised forms of transport. However, a discussion on consumption patterns and lifestyles is missing from the report.

Inclusion of organisational aspects and "social innovations"

While the aforementioned theories, research areas, and programmes all promote primarily technical solutions to environmental problems that are entirely compatible with the paradigm of economic growth and have only in part come to acknowledge the importance of the consumption side recently, it should be noted that they all also contain many organisational, institutional, or system-related aspects, as well as certain "social innovations" (see section 3.2.5). These cannot be categorised adequately under the headline of "technology" alone.

For the example of modern, customised forms of mobility, Kemp notes that changes occurring on the system level involve many innovations, including

> *"[...] innovations in vehicle technology such as urban cars, long distance cars, changes in ownership, better public transport, mobility centers where you can change from one type of transport to another, and mobility agencies who supervise and manage fleets of cars, busses, bicycles. Technology plays a role in it but it is very much an organisational issue."* (Kemp 2010: 15)

According to Kemp, system innovations "are fundamental changes in the systems of provision, in which we have change at multiple points (in technology, products, practices, organisation, ways of thinking)" (2010: 5). Thus, at least for far-reaching, systemic innovations aimed at advancing sustainability, technical developments often do not stand alone, but are combined and overlap with changes in behaviour and daily practices. This is also what is meant by "social innovations" (cf. section 3.2.5): even if mainly based on new (and a revival of old) social practices, they could often not be realised or would be less attractive without certain technological advances, as in the case of modern carsharing schemes that heavily rely on the use of smartphones and GPS availability.

To sum up, within almost all the discourses mentioned that used to focus on technical solutions, a tendency to include the perspective of consumption can be observed, even if they do not explicitly (or explicitly not) target the amount or levels of consumption, but rather consider improvements to what is being consumed. However, sometimes even advocates of technical solutions mention in passing that this may not be sufficient to achieve sustainability and that changes that affect consumption levels may be required. For instance, the authors of *Factor Five* note in the final chapter of the book:

> *"Our civilizations worldwide have to face sufficiency, or some limits to greed, growth and consumption. [...] we suggest that in addition to this* [increased resource productivity] *a movement towards sufficiency—that is, shifting lifestyles closer to the level of requirement for a decent living—will be needed."* (Weizsäcker et al. 2009: 346).

3.2 Sufficiency

As opposed to technological approaches, sufficiency deals with consumption patterns and behaviour change, and implies that there is an upper limit ("enough") to total and per capita resource consumption[132] and emissions, beyond which the environmental carrying capacity of the planet would be exceeded.

3.2.1 History of the idea

The idea that people, especially in industrialised countries, should limit their consumption because of personal, social, spiritual or also ecological reasons is of course much older than the term "sufficiency". The following overview of the history of this idea before the term was coined highlights some important references, without claiming to be complete.

Criticism of certain consumption practices is already present in the ancient Greek term "pleonexia", which can be translated as "greediness" or "graspingness". "Pleonexia" is seen, for example, by Aristotle in his Nicomachean Ethics as a main source of injustice (Nicomachean Ethics V 2 1129b 1 ff.). In the literature on sufficiency, an Old Greek phrase is also often cited, which is supposed to have been an inscription at the old Apollo Temple in Delphi[133]: *Medèn ágan*, which can be roughly translated as "not too much of anything" (Scherhorn 2002: 16, Linz 2006: 7). Thus, for some in ancient times, voluntary moderation and asceticism were seen as a virtue rather than irrational behaviour (cf. Hösle 1991: 78 f.). Afterwards,

132 On the term "consumption", see Box 5 in section 3.2.5.

133 Besides the more well-known inscription "know thyself".

for a long time the virtues of material simplicity found their place mainly within the world's major religions[134].

A systematic critique of "excessive" wealth and consumption was only brought forward much later after industrialisation and mass production had eased access to (relatively) cheap consumer goods. Early and iconic works in this regard include *Walden* by Henry David Thoreau, published in 1854 (Thoreau 1964), Thorstein Veblen's observations regarding "conspicuous consumption" in his *Theory of the Leisure Class* of 1899 (Veblen 2007), Richard B. Gregg's pamphlet *The Value of Voluntary Simplicity* (Gregg 1936), John Kenneth Galbraith's analysis of *The Affluent Society* (Galbraith 1958 a)[135], and *To Have or to Be* by Erich Fromm (1976). All these were not or, like Thoreau's and Fromm's work, were only partly motivated by a concern for the overexploitation of the environment.

From the 1970 s onwards, as the idea of limited ecological carrying capacity gained more attention, a number of reports and other writings were published which contained the idea that consumption should somehow also be limited because of environmental concerns. *The Limits to Growth* (Meadows et al. 1972) addressed the limited capacity of the planet's ecosystems to accommodate a growing population, emissions, increasing resource depletion and industrial modes of production. In other words, it provided an overall analysis of "how much is too much". Unlike one may expect, the question of "how much (individual) consumption is enough" was not directly addressed in the report. Nonetheless, and notwithstanding the criticism of the report and its projections, it is an important milestone

134 For a more comprehensive overview of the virtue of material simplicity in ancient Greek culture as well as in various religions, see for instance Voget-Kleschin (2012: 77 ff.).

135 Cf. also Galbraith's essay *How much should a country consume?* (1958 b), where he points out the impossibility of further neglecting what he calls the "conservationist's approach" to consumption of materials: "If we are concerned about our great appetite for materials, it is plausible to seek to increase the supply, to decrease waste, to make better use of the stocks that are available, and to develop substitutes. But what of the appetite itself? Surely this is the ultimate source of the problem. If it continues its geometric course, will it not one day have to be restrained? Yet in the literature of the resource problem this is the forbidden question. Over it hangs a nearly total silence. It is as though, in the discussion of the chance for avoiding automobile accidents, we agree not to make any mention of speed!"

in the consumption discussion, establishing the point that there are limits to material growth of anything within a constrained system.

Shortly after *Limits to Growth* and still more than a decade before the rise of the term "sustainability", Schumacher wrote in *Small is Beautiful* that new lifestyles and patterns of consumption need to evolve—lifestyles "designed for permanence" (Schumacher 1973: 17). He also literally mentions the problem of "enough":

> *"Is there enough to go round? Immediately we encounter a serious difficulty: What is 'enough'? Who can tell us? Certainly not the economist who pursues 'economic growth' as the highest of all values and therefore has no concept of 'enough.' There are poor societies which have too little; but where is the rich society that says: 'Halt! We have enough'? There is none."* (Schumacher 1973: 21)

Later that decade, the idea of "voluntary simplicity" was revitalised in an article by Duane Elgin and Arnold Mitchell (Elgin/Mitchell 1977). Elgin and Mitchell went beyond Gregg's earlier critique and did not only describe the advantages of a simple lifestyle for "personal growth" (Gregg: "personal hygiene") and social aspects, but also for global resource consumption and environmental degradation:

> *"A uniquely modern aspect of voluntary simplicity is that this way of life seems to be driven by a sense of urgency and social responsibility that scarcely existed ten or fifteen years ago. This sense of urgency seems to derive from many serious social problems, including: the prospects of a chronic energy shortage; [...] growing demands of the less developed nations for a more equitable share of the world's resources; the prospect that before we run out of resources on any absolute basis we may poison ourselves to death with environmental contaminants [...]."* (Elgin/Mitchell 1977: 2-3)

They emphasise that these social problems would sooner or later push society towards adopting lifestyles that correspond to voluntary simplicity, ultimately leaving no other option (Elgin/Mitchell 1977: 13).

In the early 1970 s, the idea also started to appear in political discussions and documents. The *Cocoyoc Declaration* (UNCTAD/UNEP 1974), the final document of a joint conference by UNCTAD and UNEP on environmental issues and development strategies combined the worries about underdevelopment with those about high levels of consumption. The declaration mentioned not only the social disadvantages of accelerating production and consumption cycles ("an ever-increasing need for tranquilizers and mental hospitals"), but especially pointed out the limited carrying capacity of the biosphere and resulting environmental problems.

The Brundlandt report *Our Common Future*, which followed in the 1980 s, did not deal comprehensively with the sufficiency idea, but it contained some—very general—references to the subject (e.g. WCED 1987: Overview/29, Annex 2/The Commission's Mandate). All in all, the report was, however, rather committed to maintaining the idea of economic growth (cf. section 2.3.5).

In contrast, *Agenda 21* is much more specific on this. It contains a complete chapter on "Changing Consumption Patterns" (UNCED 1992 a, Chapter 4), and the issue is addressed in many other parts on specific subjects such as energy, waste, and education. The agenda does not only highlight the global impact of unsustainable consumption and production patterns, particularly in industrialised countries. It also contains a whole programme area on "Developing national policies and strategies to encourage changes in unsustainable consumption patterns". However, even if governments are asked to "develop a domestic policy framework that will encourage a shift to more sustainable patterns of production and consumption" (UNCED 1992 a, Section 4.17), the agenda focuses on efficient use of resources and encouraging consumers to consider environmental aspects in their purchasing decisions. It does not explicitly mention the need to consume *less*.

3.2.2 The term "sufficiency" and its interpretation

The term "sufficiency" was used first by Daly in the context of the debate on economic growth (Daly 1977, 1993, 1996; cf. section 2.3.5). It was introduced into the German-language environmental discourse and contrasted with efficiency in an article by sociologist Wolfgang Sachs (1993; cf. Kleinhückelkotten 2005: 55, Winterfeld 2002: 27). In this article, Sachs stated that deploying efficiency, that is, using the available means in a more rational way, only helps to save time but does not render obsolete the need to confine the aims prudently. Thus, he concluded that "the 'efficiency revolution' remains blind if it is not accompanied by a 'sufficiency revolution'. After all, nothing is as irrational as heading in the wrong direction with a maximum of efficiency" (Sachs 1993: 69).[136]

136 My translation.

The term itself and the discussion behind it are very controversial, and manifold interpretations, rationales, and fields of application have been suggested. In the following, I will try to provide a clearer picture of the idea along a number of distinctions. Afterwards, this study will establish its own conception of sufficiency appropriate to its research focus and theoretical foundation.

Positive versus negative understanding

Opponents of the sufficiency idea hold a negative interpretation of the term as a last option when nothing else works. They frame it in notions such as frugality, renunciation, or even ascetisism (e.g. Huber 1995 b: 40, 1999: 1 f., Alcott 2008: 771, Deutscher Bundestag 2013: 715 f.[137]; cf. also Linz 2002: 9, Winterfeld 2007: 48). In this sense, sufficiency means "doing without" (Huber 2000: 275) and implies a decrease in utility (Alcott 2008: 771).

Sufficiency advocates deny this view by the majority and relate the idea to more positive notions, such as moderation, satisfaction, slowing down, freeing oneself from stress and sensory overload, "enoughness", the good life, and quality of life (e.g. Hennicke 2002: 60 f., Paech 2011, Princen 2005: 6, Linz 2002: 12, 2004, Linz/Kristof 2007, Ott/Döring 2007: 58, Scherhorn 2002: 16, Schneidewind/Zahrnt 2014: 13). They also deny that sufficiency necessarily implies lower utility and hold that utility might sometimes even be higher. This is, according to Linz and Kristof (2007: 182), the case when not only the available goods but the overall balance, also with regard to the available time and relationships, is considered. In line with this, Paech argues that there is a "critical limit" to the number of consumption activities within the available time frame, beyond which the total utility would decrease with an additional unit of consumption. Sufficiency then would mean limiting consumption activities accordingly and giving things more time (Paech 2009: 73, 78). Fischer and Grießhammer propose a definition of sufficiency that implies changes in the various aspects of utility of the consumption activities targeted—whether these

---

137 This report by the German Bundestag's "Enquete-Kommission" contains a section outlining a dissenting opinion by some of the commission's members on the subject of sufficiency, which draws a thoroughly negative (and very simplistic) picture of it, including that the idea is based on a "negative idea of man".

changes are perceived as renunciation or not, depends on the individual (Fischer/Grießhammer 2013).

Narrow and broad meaning

Linz (2002, 2004) and Voget (2009) propose distinguishing between a narrow and a broad meaning of sufficiency. The narrow meaning suggests a quantitative reduction in the use of resources for individual consumption as the counterpart to efficiency (Linz 2002: 13, Voget 2009: 209, 213). The broad meaning implies a whole new understanding of welfare, with a focus on quality of life rather than living standards (Voget 2009: 221), and a value change within society, which would also require social and political conditions that allow for these changes (Linz 2004: 11).

According to Linz (2002: 13), the narrow and the broad views are linked in such a way that the former relies on the latter, because reduced consumption will not be achieved without changes in the current way of life and performing economic activities. Some have proposed still narrower definitions of sufficiency that exclude specific circumstances of consumption reduction (cf. Alcott 2008: 771, Müller 2009: 85 f.).

Motivation: environmental carrying capacity and the good life

The reasons for promoting sufficiency are twofold: the main motivation is a concern for ecological overshoot, and a second is the pursuit of the good life or happiness. Many advocates of sufficiency base their ideas on both motivations (e.g. BUND et al. 2008: 232 ff., Müller 2009: 86)[138].

---

138 Cf. the title of a recent issue of *Politische Ökologie* (oekom 2013, my translation): "The right measure: sufficiency as a key to more happiness and environmental protection". Cf. also the following definition of an "energy sufficient life", which contains both motivations, as proposed by Müller (2009: 86): "a western urban middle or upper-class individual lives an energy sufficient life if a) s/he consumes energy not beyond some levels of 'enough'; and these consumption levels are determined by b) some understanding of the externalities from energy consumption; by c) some understanding of this individual of what s/he truly 'needs' to live a 'good life'; and d) a sense of responsibility in the individual for the effects of his/her own actions (in particular including the global externalities

The first concern goes back to the debate on the limits to growth and that there must be upper limits to the "scale" of the economy as a subsystem of the ecosystem (Daly 1996: 6 f.; Daly et al. 2007; cf. section 2.3.5). Later, it was not so much the scale of the economy but more directly the limits of environmental carrying capacity that became the focus of the debate on sufficiency. This is accompanied by a concern for justice in the face of the "Northern oligarchic living standards" (Ott/Döring 2007: 43), the globalisation of which would have devastating consequences because of their externalities. The focus from a purely environmental point of view is therefore not on consumption in general but on those acts of consumption which place a relatively heavy burden on the environment. It is thus the "composition of consumption" (Ehrlich/Goulder 2007) that is seen as problematic from the point of view of the environmental motivation.

Besides the problem of ecological carrying capacity, a number of pleas for sufficiency are also motivated by the observation that quality of life or happiness does not necessarily increase with the level of material wealth after crossing a certain threshold (Linz/Kristof 2007, Ott/Döring 2007: 51, Alexander 2012, Schneidewind/Zahrnt 2014).

Both motivations are reflected in a proposal to distinguish between qualitative and quantitative sufficiency (cf. Darby 2007: 111). The qualitative understanding is rather subjective and, according to the author Sarah Darby, implies "wealth and plenty", "something to be celebrated", and that some optimal state has been reached. The quantitative definition is more objective, uses absolute points of reference and implies a "threshold of acceptability". As proposed by Darby, quantitative sufficiency relates to bottom lines ("enough for a necessary purpose") as well as ceilings ("too much for safety or welfare"; see next section).

"Enough" as an upper and lower threshold

Related to the point before, sufficiency or the notion of "enough" can be conceived of as an upper or a lower threshold[139]. The upper threshold, which is addressed by concerns for the environmental carrying capacity, marks a point beyond which there is "too much" of something—say, con-

from consumption decisions, which traditionally are not in the focus of responsibility for the effects of one's actions)."

139 Cf. section 3.2.5 on sufficiency as a notion in justice theory.

sumption, pollution, logging, etc. The lower threshold is referred to when it comes to the minimum material wealth required to meet someone's needs[140] or for a good life (cf. Daly 1996: 220). In addition, some authors also find that there is an upper threshold to material wealth, beyond which a person does not only not get happier but happiness in fact decreases, as is claimed by some. This may be the case if, for instance, the goods possessed become inconvenient as such, or they become distressing as the owner does not manage to use them because of a lack of time, or a certain level of income can only be maintained by working overtime in a stressful job (cf. Paech 2009, Alexander 2012: 10 f., Schneidewind/Zahrnt 2014).

Most authors writing on sufficiency are in particular concerned with the upper level, as their focus of analysis is on wealthy countries, where the vast majority is perceived to be well beyond the lower threshold. Some other authors do not focus on either of these thresholds but on staying in between[141] (e.g. Boulanger 2010: 5, Darby 2007: 111), that is, not to consume too much and not too little, but to find "the right measure"[142].

The macro and the micro perspectives

Daly (1993, 1996) establishes sufficiency as both a macro and microeconomic concept. While the microeconomic perspective relates to individual emissions or resource consumption, the macroeconomic perspective considers entire nations or even the whole world[143]. Most advocates of sufficiency are concerned primarily with the perspective on individuals or households, discussing how the adverse environmental impacts of individual consumption can be decreased. However, the macroeconomic perspec-

---

140 Cf. section 2.3.3, Box 1 on "Needs".

141 A nice visualisation of this idea has been proposed in a working paper by Oxfam (Raworth 2012), which sets lower and upper thresholds in the form of a Doughnut. The inner circle (marking the hole in the Doughnut) represents various dimensions of the social foundation, while the outer circle represents various dimensions of an environmental ceiling. The paper asks: "Can we live within the Doughnut?" A weakness of this model, however, is that the scale and dimensions for the inner and outer thresholds are not identical and thus cannot be compared.

142 *"Das rechte Maß"*—a very common phrase in the German-language literature on this subject.

143 Cf. also the discussion on the term "consumption" below (Box 5, section 3.2.5), which features the same two perspectives.

tive is usually implicit in this and it seems to make most sense to combine both: the impacts of individual consumption aggregate at the macroeconomic level into overall environmental impacts; if these are to be limited in line with environmental carrying capacity, overall limits must be set that can be scaled down again in order to estimate sustainable per-capita levels of resource consumption and emissions.

The absolute versus the relative

As Müller (2009: 83) states,

> *"[s]ufficiency [...] addresses the level of output (or consumption) per se—and not in relation to the inputs. It asks whether an activity needs to be performed at all (resp. at a certain level) and not whether it is performed 'efficiently' (resp. at an efficient level)."*

This is probably shared by most sufficiency advocates. However, some frame sufficiency as an idea that does not exactly aim at reducing consumption or its externalities to a certain level, but only leads to a relative decrease compared to the level that existed before (cf. UBA 2012: 36).

Moreover, there are proposals to frame sufficiency as an optimisation problem with regard to the material resources or time used for consumption. For instance, Weizsäcker et al. state that "satisfaction per service" is a measure of sufficiency (2009: 347): "As people enjoy more satisfaction from the same service, their felt quality of life is increasing without added resource consumption and even without added efficiency." In a similar way, Boulanger (2010: 5) suggests two sufficiency attitudes, the first of which is "striving to get the maximum wellbeing from each unit of material service consumed". Both the proposals of Weizsäcker et al. and Boulanger would, however, not automatically lead to a decreased level of consumption, as someone could still consume at a very high level and claim to be much more satisfied with each new television and car he or she buys.

The idea could still be rescued by Paech, who introduces an absolute parameter by referring to the limited time one has for one's consumption activities (see above). He states that after someone has used up all their available time, the addition of one more consumption activity would decrease the utility of the goods already acquired more than the new activity can add in utility (Paech 2009: 73). However, such a "critical limit" is, again, highly subjective and cannot make sure that overall consumption is fixed at a moderate level. In addition, one can make use of time-saving technologies or services (that might be highly polluting) to extend one's available time. Also, any view of sufficiency framed in terms of "consumption efficiency" entails the problem that it causes further confusion with regard to its distinction from the technical strategy of efficiency.[144]

Boulanger adds a second sufficiency attitude to his first, which stipulates "minimizing the role of material services in the definition and production of wellbeing (cultural dematerialization)" (2010: 5). This is more in line with what most sufficiency advocates have in mind, but it is still an optimisation problem, so that it is not clear *at which level* minimisation will be completed. It could be the level of material consumption that is absolutely needed to survive, which might be below what ecological carrying capacity can endure. In this case, it is unclear why material consumption should be reduced that far. Thus, it can be concluded that sufficiency does not work as a purely relative concept, but has at least to be accompanied by some idea about the sustainable level of consumption. This also makes sense as the term sufficiency is derived from the notion of "enough", which by its very meaning is an absolute concept (see above "enough as upper and lower threshold").

Voluntariness and dependence on external structures

It is usually emphasised in the literature on this subject that sufficiency, with its elements of modesty and limitation, should mainly be adopted voluntarily and not due to poverty and misery or coercion (Bartelmus 2002: 40, Linz 2002: 13, Ott/Döring 2007: 58, Alcott 2008: 771, Stengel 2011: 140). Thus, most confine their claims for sufficiency to people in rich countries, or even wealthy people in rich countries (e.g. Müller 2009: 85). The exclusion of (non-voluntary) poverty as a driver for adopting a lifestyle characterised by sufficiency is undisputed. This is also reflected in the idea that sufficiency could be motivated by some account of the good life, which excludes non-voluntary poverty and misery.

However, at the same time many authors stress that individual lifestyles do not exist independently of the surrounding conditions and thus empha-

144 Compare Goodland and Daly's definition of sufficiency as "doing more with less" (1996: 1009), which quite obviously describes the idea of efficiency.

sise the need for appropriate social, economic and political structures or conditions in order to be able to adopt lifestyles compatible with sufficiency (e.g. Hennicke 2002, Winterfeld 1993, 2007, Ott/Döring 2007: 41). This issue will be dealt with more in detail later with regard to a potential policy of sufficiency (cf. section 4.1).

Reducing not only consumption but also population size?

In some cases, sufficiency is explained as a variable of the famous "IPAT" formula (e.g. Huber 1999: 1, Alcott 2008: 773, Weizsäcker et al. 2009: 348). This formula, once presented by ecologist Paul R. Ehrlich and colleagues[145], describes environmental impact as the product of three factors:

Impact = Population × Affluence × Technology

Sufficiency is usually seen as a strategy to reduce A, that is, the level of affluence or at least that of resource-consuming affluence (Weizsäcker et al. 2009: 348). However, it could in addition relate to the population size. While sufficiency demands changes in the consumption patterns in rich economies, for developing countries it might be seen as a strategy for population control (Huber 1999: 1 f., 2000: 275).

In this sense, Daly has noted that in a "full world", the human population and the level of affluence (consisting in the amounts of cars, buildings, livestock, etc.) compete with each other, as more of one comes at the expense of the other (Daly 1996: 119). He gives priority to a higher number of people and proposes that what should be limited first in terms of sufficiency is material wealth (1996: 220; cf. also Daly 1993: 361): "We should strive for sufficient per capita wealth—efficiently maintained and allocated, and equitably distributed—for the maximum number of people that can be sustained over time under these conditions." This seems to imply that limiting the number of people, in a second step, is also an option, if necessary.

Most authors writing on sufficiency, however, are silent on this issue as they conceive of sufficiency as a strategy solely for affluent countries, which usually do not (or only insignificantly) have growing populations.

---

145 For instance, in an early version as I = P × F(P), with F(P) as per-capita impact (Ehrlich/Holdren 1971).

In addition, population control is an awkward subject, so most researchers are certainly happy to leave it aside.

How to distinguish sufficiency from technical sustainability strategies?

Many texts on sufficiency struggle to find ways to clearly distinguish it from technical sustainability strategies. As already mentioned before, some authors try to use the concept of utility for this purpose. For instance, Alcott (2008: 771) suggests that sufficiency

> *"[...] is not the same as consumption efficiency, by which is meant behaviour that achieves a given level of utility with less (energy) input: e.g., boiling only the amount of water needed for the cup of coffee, switching off unneeded lights, or carpooling. [...] Sufficiency, in contrast, means doing without the cup of coffee, getting by with dimmer lighting, and not taking the car. That is, assuming that 'environmental concern' is left out of the utility function, sufficiency implies lower utility or welfare."*

However, it would not be easy in practice to distinguish between a light being switched off because it is not needed and the same action leading to "lower utility". Similarly, while for one person doing without the cup of coffee might result in a bad mood and indeed much lower utility, for the other, even when excluding environmental concerns, the decision might result in higher utility in terms of health. Moreover, a decision for or against a cup of coffee depends on the number of cups one has had before. The same can be said for not going by car and going by bike instead, for instance. Maybe what Alcott wants to say is that *if* there is a lower level of utility in an action related to environmentally sound consumption behaviour, *then* this is what he labels sufficiency. In that case, however, it is unclear why people should behave that way at all, since it is assumed (in Alcott's analysis) that they do this voluntarily. Finally, such a highly subjective concept cannot really serve for analytical purposes.

What is more convincing is the proposal by Fischer and Grießhammer (already mentioned above) that sufficiency implies changes in the various aspects of utility that are related to the consumption actions in question—whether these changes are perceived as a decrease in utility or not depends, again, on the individual (Fischer/Grießhammer 2013). In contrast, technical solutions, according to them, aim at supplying exactly the same utility at a lower level of environmental impact. Whether this distinction is as clear as they claim (Fischer/Grießhammer 2013: 10 f.) is, however, questionable. Does changing a traditional incandescent lamp for an energy efficient fluorescent lamp provide for the same utility, even if I am not entirely satisfied with the luminous spectrum of the lamp? This is, again highly subjective, so it would be impossible in many cases to label an improvement in environmental performance as a mere technical change or not.

Any other attempts to distinguish sufficiency from technical approaches are rather descriptive and do not rely on a single criterion such as utility. One example in this regard is the claim that sufficiency involves changes in the lifestyles or behaviour of consumers, while technical approaches do not demand such changes in consumption practices, at least no manifest ones (Stengel 2011: 130). What exactly changed lifestyles or consumption habits mean, however, is not obvious. Mostly, this is described by means of examples, such as doing without one's own car, reducing meat consumption, extending the use phase of products such as furniture and clothes, or intensifying it through collective forms of use. However, how can we judge examples such as going to the eco-supermarket instead of a conventional one and buying organic food? Is this a behaviour change already, or does it rely on the (rather) technical approach to changing production methods? Apart from examples in more abstract terms, various ideas exist to describe the desired consumption behaviour: some understand sufficiency as lifestyles characterised generally by a low level of consumption (Voget 2009: 211 ff., Ott/Voget 2007); elsewhere it is stated that sufficiency aims at reducing or replacing practices that excessively consume resources (Stengel 2011: 140, Linz 2006: 6), or at replacing the purchase of goods with other processes (Voget 2009: 220). The most radical position on this is expressed by Paech, who denies any close relationship between sufficiency and "sustainable consumption". Instead, he states that sufficiency is actually the opposite of sustainable consumption, standing exclusively for "non-consumption" (Paech 2009: 63). Only this very radical position serves to distinguish sufficiency clearly from technical strategies. I doubt, however, that this is appropriate for defining sufficiency. For instance, if I give up my car and use public transport instead, I still "consume" the services of public transport. Excluding such cases would make sufficiency an inappropriately narrow notion.

Other examples of the aforementioned descriptive sense include equating sufficiency with "social innovations" (cf. section 3.2.5) as opposed to technical innovations; linking sufficiency to "subjects", that is, individuals, lifestyles, or even countries, and efficiency and consistency to "objects" or, in other words, goods and services (Paech 2011)[146]; or stating that sufficiency seeks to limit the level of output or consumption, and is not concerned with its relation to inputs, unlike efficiency (Müller 2009: 83; see above). For any of these distinctions, it is easy to find examples that blur the difference claimed.

It seems that sufficiency cannot be separated along a perfectly clear line from approaches that seek to decrease environmental impacts by improving the technical characteristics of products and systems of provision. It is also unclear whether such strict differentiation is needed, as the common aim is to reduce the adverse environmental impacts of human patterns of consumption and production. Strict differentiation would only be necessary if, for instance, one aims at aggregating the achievements of either of the strategies to demonstrate that it is superior to the other, which is not the purpose of this study. Instead, I think that both approaches are needed and complement each other.

What can still be said is that sufficiency certainly has an emphasis on behavioural changes, while it sometimes involves technical innovation as well. An example of this is car sharing, which is certainly a case of sufficiency (provided that a private car is given up), but also requires some technical prerequisites to work well. Thus, it might be helpful to use a number of juxtapositions—consumption/production, demand/supply, subject/object etc.—to clarify the application area of the terms, even though they do not serve for strict separation (Table 4).

146 Paech states that objects, regardless of how efficient or environmentally sound they are, can as such never give any indication of their contribution to sustainability. Thus, he suggests that sustainability performance should be measured exclusively from a subjective perspective, for instance by means of individual environmental or carbon footprints (Paech 2011: 287 ff.; cf. section 4.3.6).

Table 4: Linking sufficiency and technical solutions with other concepts

Sufficiency	Technical solutions
Consumption[147] Sufficiency focuses on individual/household and macroeconomic consumption and their associated environmental impacts, including the impacts that are "embedded" in a final product. If the impacts from consumption are, on a global level, too high, then sufficiency demands curbing environmentally unsound consumption. While this is the main focus, there are also some contributions to the debate that link sufficiency and production (e.g. Reichel et al. 2009, Schneidewind/Palzkill-Vorbeck 2011).[148]	*Production* Technical improvements primarily address production, that is, they include the design of products that are more energy efficient and more easily recyclable or production processes optimised in terms of their environmental impacts. Note that the perspective of "sustainable consumption" (cf. section 3.2.5) also commonly takes into account purely technical strategies, such as replacing an old refrigerator with a new, more energy efficient one.
Demand Sufficiency primarily addresses demand. Any approach by an individual company to apply sufficiency to supply, that is, to produce less, would encourage other producers to fill the gap with more of their products, as long as demand remains the same. However, on a macroeconomic level, limiting overall supply with quotas, as in the case of fisheries, can be a sufficiency strategy.	*Supply* Technical strategies aim at greening supply and the goods or services supplied to consumers.
Subjects Sufficiency addresses the consumption levels of individuals, households, or other groups of human be-	*Objects* Technical solutions seek to improve the environmental characteristics of the things or objects that

147 For the term consumption, cf. Box 5, section 3.2.5.

148 Reichel et al. (2009) intend to link sufficiency, consumption and production in collaborative utility systems such as car sharing. Schneidewind and Palzkill-Vorbeck (2011) list this and similar approaches, which are, however, all linked in some way to consumption.

Sufficiency	Technical solutions
ings, therefore focusing on the subjects that rely on certain consumption activities for their well-being.	individuals purchase and use, as well as the production facilities to manufacture these goods.
Behaviour change Sufficiency usually implies behaviour changes for the individuals concerned.	*Technical change* Advances in technology often do not require changes in behaviour. Systemic technical innovations, however, may also involve behavioural (in the sense of adapting behaviour to the changed technical circumstances) and in particular organisational changes.
Absolute limits Sufficiency aims to respect absolute environmental limits and asks for absolute reduction in environmental impact rather than relative improvements of input-output ratios.	*Relative improvements* As technical solutions focus on objects such as household appliances or industry plants, the environmental improvement brought about by them is relative to the previous status quo. They are usually not linked to overall environmental impacts.

3.2.3 Sufficiency in this study

The preceding discussion has demonstrated that a diversity of interpretations of the term "sufficiency" exist in academic literature. Each of them highlights different aspects, depending on the purpose and context in which it is used. The more deeply one enters the discussion on sufficiency, the messier the term appears to become. However, I think that by analysing the different facets of the term and "surrounding" it this way, it is still possible to develop a clearer understanding of it. Following the analysis above, sufficiency as used in this study can be summarised as follows:

- Sufficiency is connoted *neither positively nor negatively*, but rather seen as a neutral necessity; it might in some cases be perceived by some individuals as an increase in utility, and in other cases or by other individuals as a decrease—and utility may also remain equal for yet other individuals.

- A broad meaning would imply comprehensive social changes, including a new understanding of welfare and a change in values. Here, a *narrow* meaning, focusing on consumption levels, is used in order to make the term applicable to this context.
- The *motivation* to apply sufficiency is to stay within the *limits of environmental carrying capacity* rather than some account of the good life. The reasons for this will be explained later on (cf. sections 4.2.1 and 4.4.1). Sufficiency is thus understood as a notion aimed at limiting consumption acts that entail a relatively heavy environmental burden, rather than limiting consumption as such.
- Sufficiency refers to an *upper threshold of "enough"*; nevertheless, within the broader framework of sustainability the idea of sufficiency is also linked to considerations of justice that aim to allow other human beings to achieve some lower threshold of "enough", or in other words, meet their needs (cf. sections 2.3.3, 3.2.5).
- Sufficiency should be related to both on the macroeconomic (e.g. national) as well as on the microeconomic (individual) level.
- As stated above, sufficiency does not make sense as a purely relative notion. Thus, it acquires an *absolute* dimension by linking it to sustainability and environmental carrying capacity.
- While it is stressed that voluntary sufficiency practices are important, the focus of the analysis that follows later (cf. chapter 4) is on the *political level and dependence of individuals on outer structures*, which means that sufficiency will be explored primarily from this perspective.
- As I conceive of sufficiency as a sustainability strategy for the wealthy part of the world, which is not confronted with significant population growth, *only the level of consumption* will be in focus, and not that of the population.
- As regards its *distinction from technical sustainability strategies*, I doubt the possibility of finding any simple criterion or form of description that clearly provides for such a distinction. Also, I think that clear separation is not necessary, and that a definition that allows for a rough-and-ready correlation is good enough.

The intention here is therefore not to define sufficiency strictly in order to separate it clearly from non-sufficiency cases (unlike e.g Alcott 2008 and Müller 2009), but to use a simple definition which, first, works for the purpose of the analysis here, that is, it should be applicable to the political level although it can be used with regard to individuals as well; second, that contains a reference to the purpose of staying within the environmental carrying capacity; and third, that avoids the impression that reducing consumption is seen as an end in itself that should be pursued infinitely, which requires referring only to environmentally significant consumption and establishing a link to environmentally sustainable per-capita consumption levels:

Sufficiency is the reduction of demand or use of goods and services with high environmental impacts in order to achieve per-capita consumption levels that allow for staying within the environmental carrying capacity.

This definition thus does not specify whether an activity is not performed at all or is replaced by a similar, less resource intensive activity to achieve the same aim, such as riding a bike instead of driving a car. Hence, consumption patterns in this sense may shift to less environmentally significant goods or services.

3.2.4 Limitations and obstacles to sufficiency as a sustainability strategy

Generally, and compared to technical measures, sufficiency is assumed to be effective in achieving the aim of reduced environmental impact. However, there are also limitations to this strategy and obstacles that impede its realisation, which are discussed in the following.

Macroeconomic rebound effects

Advocates of the sufficiency strategy often forget that a well-known flaw of the efficiency approach, macroeconomic rebound effects, can also occur when only a number of households, cities, or nations reduces its energy or resource consumption. The effect is the same in the end, as Alcott has pointed out: “In constituting a drop in demand, it [frugal behaviour] initially lowers prices, and this in turn raises others’ demand, so that in the end some of what was ‘saved’ through non-consumption is consumed after all—merely by others.” (Alcott 2008: 775). Alcott qualifies this statement by saying that this analysis holds for the consumption of raw materials, such as metals and energy sources, but it remains open whether it also applies to other consumption items such as food, water, or clothing (Alcott 2008: 775).

With regard to microeconomic rebound effects, and in contrast to efficiency, here it is assumed that for the sufficiency strategy these effects will not occur when a household, city etc. voluntarily reduces its resource consumption for the purpose of lowering environmental impacts. This assumption seems justified, because if a household aims for an absolute reduction in resource consumption, it would not compensate for this reduction through other actions that would lead to an equal amount of resources being consumed.

In conclusion, for some resources reduction strategies may not be successful if not implemented at the highest possible level (worldwide, nationwide etc.) in order to avoid macroeconomic rebound effects. Still, depending on the type of resource and environmental impact involved, sufficiency can generally be beneficial on local or regional levels, for instance in the form of fewer cars that result in reduced local air pollution and noise and more free space (Alcott 2008: 782). In addition, a reduction strategy can also be an example to others and have encouraging effects on them (Heyen et al 2013: 18).

Acceptance and other obstacles

Some authors have expressed their doubts about the potential of sufficiency to be practised by a number of people large enough to have an impact, therefore questioning the overall effectiveness of the strategy (cf. Kraemer 1997, Huber 1995 a, 1999, 2000). Stengel (2011: 183 ff.) has carried out a comprehensive analysis of the obstacles to sufficiency being accepted and implemented. He identified five main barriers:

- Personal costs involved, such as convenience, social status, habits, time etc., when lifestyles change.
- Consumerism: due to the prevalent materialistic self-perception among people, visions of the good life and social acknowledgement are widely linked to the consumption of goods and options. Only a limited number of people within alternative social milieus feel attracted by a "culture of renunciation" (cf. also Huber 2000: 275, Kraemer 1997: 201).
- Social conventions: the practice of the majority is closely tied to prevalent opinions, trends and conventions. As long as sufficiency practices are not adopted by a critical number of people, the majority will not consider them attractive.

- Uncertainty about the allocation of responsibility: at the political level there is widespread agreement that consumer sovereignty has priority over any measures limiting consumption and that any sufficiency approach on a political level would be clearly reflected in the next elections. However, individuals often have the impression that their decisions have no influence and that therefore politicians and the industry are responsible for environmental issues.[149] This uncertainty can also be explained by the probem that the main benefits of sufficiency, positive effects on the environment, are socialised and of medium to long-term in character, while many individuals primarily perceive acting in line with sufficiency as an immediate constraint on their own lives (cf. Heyen et al. 2013: 18 f.).
- Capitalism and the growth economy: the current economic system requires growth and therefore encourages consumption through diverse mechanisms. This system continuously creates new consumption options and devaluates old ones.

To this last barrier it should be added that the current economic system does not only encourage consumption, but that there is also significant uncertainty about the potential effects on the economy and the labour market if sufficiency practices were adopted by a relevant number of people (cf. Linz 2006, Heyen et al. 2013: 18 f.). The effects are likely to be different for different sectors of the economy—those demanding a high input of resources in their production processes will be affected more strongly, while others such as labour-intensive service sectors might even experience further growth. Similar discussions are held within the "beyond growth" debate (cf. section 2.3.5). However, whether sufficiency will affect economic growth is left open here, as it is at least in theory possible that decreases in demand in some sectors are compensated for by shifts to other sectors.

Finally, with regard to technology development, it was stated above that this requires time and comes at a not insignificant cost. Consumption practices can, in principle, be changed from one day to the next (Stengel 2011: 148 f.), assuming the willingness of consumers and disregarding the social and economic consequences as well as the required organisational adaptions. However, changes in lifestyles on a society-wide level, if they are to happen based on a transition in values and dominant views of the

149 Cf. sections 4.1.1 and 4.1.4.

good life, of course also require considerable time (Brook/Bradshaw 2012).

3.2.5 Related concepts and debates

There are a number of concepts that (at least at first sight) are similar or feature certain linkages to sufficiency, as their use in the literature on this subject suggests. In order to clarify possible links and overlaps, I compare sufficiency to and distinguish it from these concepts in the following.

Sufficiency as a notion in justice theory

The term "sufficiency" is also used in a context quite different from that outlined so far, namely within the philosophical discourse on justice and equality[150]. In terms of literature references, these two sufficiency debates are not directly linked to each other (Spengler 2016 a). Discussions about sufficiency within justice theory, at least insofar as the term sufficiency is concerned, were launched by an article by Harry Frankfurt (1987). In this article, he proposed a "Doctrine of Sufficiency" (1987: 21 f.):

> *"With respect to the distribution of economic assets, what is important from the point of view of morality is not that everyone should have the same but that each should have enough. If everyone had enough, it would be of no moral consequence whether some had more than others. I shall refer to this alternative to egalitarianism—everyone to have enough—as 'the doctrine of sufficiency.'"*

As indicated in the quotation, advocates of sufficientarian justice[151] are concerned with some level of "enough". According to them, the achievement of justice hinges upon everyone achieving this threshold. Above the threshold, it does not matter how goods or income are distributed. This is

---

150 Cf. the discussion on absolute standards in section 2.3.3. See in particular Frankfurt (1987, 1997), Crisp (2003), Casal (2007), Page (2007 a), Meyer (2009).

151 In order to distinguish the two sufficiency concepts from each other, in this section I will refer to the one developed in philosophy as "sufficientarian justice" (the term "sufficientarianism" is commonly used as a label for this position), and to the one common in environmental discourse, with which this study deals, as "environmentalist sufficiency". Elsewhere in this study, I always refer to the environmentalist concept with the term "sufficiency".

a characteristic sufficientarian justice has in common with environmentalist sufficiency—both aim at establishing some level of "enough" (or a level that is "sufficient"). As was already pointed out in section 3.2.2, there can be lower and upper levels of enough. As opposed to sufficientarian justice, the environmentalist concept of sufficiency primarily focuses on the upper level[152].

Sufficientarian justice is, at first glance, not concerned with adverse environmental impacts, but with material—and sometimes immaterial[153]—goods or resources required for human well-being. However, it is indirectly linked to environmental issues, for instance when it comes to possible duties current generations have with regard to environmental protection for the sake of future generations (cf. Page 2007 a). Material wealth is, in turn, of interest within the environmentalist sufficiency concept only insofar as it produces adverse impacts on the environment. An individual's virtual amount of money in a bank account, even if it is enormous, would as such not matter to the environmentalist. Environmentalist sufficiency, therefore, has to do with distributive justice only insofar as the distribution entails harmful emissions and the overexploitation of natural resources.

Satisficing

"Satisficing", in the meaning coined by the economist and psychologist Herbert A. Simon, is a behavioural strategy in microeconomics. The traditional strategy that economists expect individuals to adopt is "maximising" or "optimising". This strategy rests on the assumption that people have skills (substantive knowledge, a stable system of preferences, the ability to compare all options rationally) that enable them to achieve the optimal result in terms of their preferences (Simon 1955). However, as Simon claimed, instead of maximising, human beings usually satisfice, that is, they avoid examining all possible options (or are simply not able to do so) and rather try to find a solution that is satisfactory or "good enough" (Simon 1965: xxiv ff.). This claim has triggered substantial debate, as it challenges established models in rational choice theory.

152 For a detailed comparison of both concepts, see Spengler (2016 a).

153 For instance in Martha Nussbaum's capabilities approach (2011: 20 ff.), although Nussbaum is not strictly a "sufficientarianist" as she does not use that term.

At first sight, the idea of satisficing in economic theory appears not to have much to do with sufficiency. However, the term has also been used outside its original context. Daly, for instance, uses satisficing and sufficiency as synonyms. He suggests that within a steady-state economy the level of per capita products should be "satisficed", since "beyond some level of sufficiency further increase in per capita goods does not increase quality of life" (Daly 1993: 361). Elsewhere he states that in such an economy the "stock", that is, "the total inventory of producers' goods, consumers' goods, and human bodies" (Daly 1991: 35), has to be "satisficed" —which means that it is to be kept at an ecologically sustainable level that is sufficient for an abundant life for the present generation and long into the future (Daly 1993: 328, 1991: 94).

In a similar vein, but rather contrasting the two terms, Princen has picked some quotes from Simon's work and put them into his own context of sufficiency (Princen 2005: 18):

> *"[...] I use sufficiency to mean a sense of 'enoughness' and 'too muchness,' a quality where concern for excess is paramount in the life of an individual, an organization, or a nation. But in everyday English the term often means or implies 'good enough' or 'second best'. In economics, psychology, and other fields, satisficing appears to have similar connotations.* 'Human beings ... satisfice because they have not the wits to maximize,' *wrote psychologist Herbert A. Simon in a work that landed him a Nobel Prize in economics.* 'Administrative man,' *as opposed to the superrational* 'economic man,' *satisfices* 'because he treats the world as rather empty and ignores the interrelatedness of all things...,' *Simon wrote. My use of sufficiency could not be more contrary."*

Princen thus equates Simon's notion of satisficing with a negative understanding of sufficiency. What Princen overlooks is that his use of sufficiency and Simon's use of satisficing cannot simply be compared. Simon just tried to show that individuals *do*, in fact, usually satisfice and that this is a rational strategy. Moreover, Simon's "maximising" does not necessarily refer to getting as many material possessions as possible, but it generally means maximising utility or achieving the best result (in terms of one's preferences). Contrarily, Princen[154] as well as many other sufficiency advocates propose a normative notion of sufficiency that aims at convincing

154 Actually, in other passages of his book Princen also describes sufficiency as a value-neutral, rational strategy of self-control, for instance in restricting oneself to eating less chocolate than one might desire in order to avoid adverse effects (Princen 2005: 6 f.).

people to reduce their material consumption for ethical reasons. Thus, sufficiency can be translated as the claim that wealthy people *should* satisfice with regard to material goods more often. This does, of course, not mean that striving for less than the maximum material wealth possible might not sometimes (or even often) be rational as well.

Nevertheless, since Simon's times the discussion on satisficing has developed further and the notion has acquired a diversity of meanings. In a volume named *Satisficing and Maximizing*, the editor notes with regard to the term satisficing that "[i]n any conversation of this complexity it is always in point to ask whether the participants are talking about the same thing." (Byron 2004: 1). Indeed, very different lines of argument have evolved around the term. Among them, satisficing appears in a normative sense as a strategy to act in the name of the virtue of "moderation" (Byron 2004: 6 ff.). In this meaning, satisficing comes closest to sufficiency.

One example in this sense is Slote's account (2004), who contrasts satisficing as intrinsic and as instrumental moderation. Intrinsic moderation is characterised by modest needs or desires, so that one does simply not want or feel the need for more of something, without any underlying considerations or reasons. As opposed to this, instrumental moderation is adopted as a strategy if someone wants to achieve a certain aim, such as not eating more in order not to gain weight, or more generally controlling one's desire for more in order not to harm oneself or others (Slote 2004: 16). This distinction is similar to the different reasons for advocating sufficiency as presented in section 3.2.2. The idea of intrinsic moderation seems to be in line with adopting sufficiency as a strategy for the good life, while instrumental moderation is akin to sufficiency as a strategy with which to stay within ecological carrying capacity and avoid harm to others.

The discussion on whether individuals (or consumers) do rather maximise or satisfice, and if they satisfice, whether they do so on intrinsic or instrumental grounds, is certainly interesting for the subject of this study. However, it seems obvious that even if a majority of individuals in wealthy countries do regularly satisfice, they do not do this on a level that would be required for environmental reasons. This is in line with Schmidtz' view, who rejects any link between Simon's concept of satisficing and the notion of moderation. Schmidtz (2004) states that while satisficing is the opposite of optimising, moderation contrasts with being immoderate, and not with optimising. Thus, satisficing does not generally imply that one's choices would be moderate, since a solution that is not optimal but "good enough" could still by far exceed the moderate level (Schmidtz 2004: 32). In contrast to such a relative understanding of satisficing, Schramme (2006: 149 ff.) proposes linking the idea to an absolute conception of basic human needs in order to ensure environmental conservation. In that case, the concept of satisficing could easily be replaced by an account of sufficiency.

What can be learned from this is that there are some parallels between the ideas of satisficing and sufficiency, and it could be useful for sufficiency advocates to look into the satisficing debate and borrow arguments from there if they intend to sharpen their analysis. This would, however, be beyond the scope of this study, so I will not go any further into this.

Consumption research and sustainable consumption

A further strand of research, which yields important contributions to the sufficiency discussion without using the term itself, has to be mentioned. This is the area of consumption research, spreading (at least) across the disciplines of economics, ecology, psychology, philosophy, sociology, and anthropology. Building on earlier pioneering works but also reflecting critically on them, research activities in this field have increased substantially since 1990, in particular with regard to the environmental or sustainability perspective.[155]

The subject of "consumption" is being debated with regard to various questions, including: What are the environmental and social consequences of consumption? Which responsibilities do consumers have as a result of these consequences? Which social functions and meanings does consumption have? What is the relationship between well-being and consumption? What are the drivers of consumption? Should policies be created in order to influence consumption patterns? Many authors, such as Tim Jackson (2005), hold that in some cases there can be "win-win" situations, where consumption is reduced and this turns out to be beneficial for both individual well-being and the environment. On the other hand, there are also

155 To give but a few references, see Durning (1992), Campbell (1994), Westra/Werhane (1998), Røpke (1999, 2010), Princen (1999), Princen et al. (2002), Sanne (2002), Shove/Warde (1998), Hertwich (2005), Jackson (2005, 2007), Schor (2005 b, 2010), Reusswig/Battaglini (2006), Soper/Trentmann (2008), Dauvergne (2010), Heidbrink et al. (2011), Reisch/Hagen (2011), Miller (2012).

voices that criticise certain tendencies to condemn consumption in an overly simplistic way, emphasising its important social and cultural functions (e.g. Miller 2012) or its inherent "good" aspects (Campbell 1994). Finally, the term "consumption" itself is subject to debate (see Box 5 below).

One might think that the idea of "sustainable consumption" corresponds, more or less, to what is meant by sufficiency. This, however, is not the case, as the debate on sustainable consumption is broader than the sufficiency debate. It also includes purchasing more energy efficient appliances or changing to a "green" energy utility company, thus taking into account technical solutions as well. Existing policies that aim at sustainable consumption mostly promote such technical solutions (Rubik et al. 2009: 127), apart from their aim of informing consumers to enable them to make better choices (cf. e.g. EC 2008, OECD 2008, Bundesregierung 2016). There are authors who go as far as saying that sufficiency is actually not a close relative of sustainable consumption, but its counterpart: non-consumption (cf. Paech 2009: 63, 2011; see also section 3.2.2). Indeed, in papers and studies by public authorities, as Jackson (2007: 256) has observed, there tends to be a consensus that "sustainable consumption means (more) consumption of more sustainable products and this is achieved primarily through improvements in the productivity with which resources are converted into economic goods", which has little to do with sufficiency. I think, however, it makes sense to say that sufficiency is, or should be seen as, a component of sustainable consumption, as any meaningful discussion on how to consume sustainably cannot avoid a consideration of consumption levels. This is also suggested by Princen, who asks consumption researchers to include "non-purchase decisions" in their analysis (2002: 28). Further, it is a part of many visions regarding sustainable consumption or how a sustainable economy should work, such as Juliet Schor's "Plenitude" (2010) or Kate Soper's "Alternative Hedonism" (2007, 2008 a, 2008 b). As I conceive of sufficiency as a part of sustainable consumption, I will use the insights from consumption research for my later analysis of sufficiency as an individual and political concept (see section 4.1).

Box 5: Consumption as a concept in environmental science

The various disciplines that deal with consumption use a broad range of definitions which focus on different aspects (cf. Stern 1997). I will only deal with the environmental perspective here, for which the focus should be on environmentally significant consumption. As indicated above, there are many other aspects of consumption, such as its social and cultural functions or its role within the economy, which are studied within other contexts and cannot be considered in detail here.

Consumption in colloquial language simply means using something up with the result that it becomes unavailable for further use. Looked at more carefully, what we colloquially refer to as consumption can mean very different kinds of activities, including purchasing, using, or using something up (cf. Princen 2002). As Rosa (2011: 128) conceives the idea of consumption, a thing is consumed only when it is being used (reading a book, knowing how to play a piano) and not when it is put on the shelf at home and simply possessed. As it is easy today to possess many more things than can be used because of the comparably low prices of commodities, consumers concentrate in reality on purchasing and not on consuming goods and services. However, if we are to address the environmentally problematic aspects of consumption, we need to include both use patterns and acts of purchasing (Princen 2002: 27), since for many goods the main environmental burden stems from the production phase.

Further, the notion of "using up" is problematic, as thermodynamics has taught us that neither matter nor energy can be used up (or produced), but can only be transformed into a less (or more) useful state. For this reason, some sustainability theorists claim that consumption as such is not problematic at all. According to McDonough, for instance, we only need to design products in a way that they either consist of organic material, so that they can be "consumed" (or transformed) and go safely back to the soil, or they consist of material that is used in the technical sphere and is only "borrowed" by the person who purchases the product and after its use the material is returned to the technical cycle. In the latter case, what people want is the function or service the good provides and not the good itself, which is why McDonough calls these "service products" as opposed to "consumption products", which one can actually use up (McDonough 1998: 305). However, other theorists do not endorse this view. Daly emphasises that in each production and recycling cycle energy is used which cannot be recycled, and some of the material is dissipated without the possibility of its retrieval. Therefore, even if we do not use up matter or energy, we still do "consume (irrevocably use up) the capacity to rearrange" matter or energy (Daly 1996: 65). Moreover, Daly resists any understanding of consumption that conceives of it merely as the reverse process of human production (adding value). Accordingly, before humans import resources into the economic system and add value to them, value has been added by nature. This capacity of nature to add value must not be overstretched, and in addition it must be considered that some consumption activities are irreversible in the sense that a similar state cannot be brought back, like in the case of the destruction of special nature sites. This view is also expressed in Stern's working definition of environmentally relevant consumption:

"Consumption consists of human and human-induced transformations of materials and energy. Consumption is environmentally important to the extent that it makes materials or energy less available for future use, moves a biophysical system toward a different state or, through its effects on those systems, threatens human health, welfare, or other things people value." (Stern 1997: 20)

The image of natural assets that are imported into the economic system leads to a further problem with regard to consumption: where does "consumption" actually occur in the production-consumption cycle? In Daly's terms, consumption of natural capital occurs when it is imported into the economic system, which occurs mainly during production processes. This is, however, not what we refer to when we speak of consumption in the sense of the "consumer society". Thus, from an environmental point of view, there are at least two different perspectives on consumption, as Røpke has pointed out (1999: 399 f.). These are, on the one hand, macroeconomic consumption of resources and "sink" capacity, and on the other final consumption by individuals or households. The macroeconomic perspective reflects Daly's view. The final consumption perspective includes all final products consumed nationally by households (or individuals) and, depending on the focus of analysis, also by the public sector. Within the perspective of final consumption, one can further distinguish between the aggregate consumption of final products and the micro-perspective of a single individual or household. Even though both the macroeconomic and the final consumption perspectives should be assessed separately, they are also interconnected, as the level of final consumption is one of the factors that drives overall resource use (Røpke 1999: 400).

To further complicate things, the environmental view on final consumption usually takes into account the whole life cycle of the goods or services consumed, including all impacts throughout the production chain, use, and disposal (OECD 2002 a, EEA 2013 a: 23). Viewed this way, the impacts of consumption cover most environmental impacts worldwide,

including, among other things, the entire production chain[156], provided that products are consumed at some point in time or have at least in principle been produced for the purpose of future consumption[157]. The difference between the perspectives of production and consumption here lies in the allocation of impacts—to a product or individual, and not to a specific production process, production site, or company.
The environmental impacts of final consumption allocated to individuals can thus be differentiated into direct and indirect "embedded" impacts (cf. EEA 2013 a: 48, Reusswig/Battaglini 2006: 5). Some consumption activities directly produce adverse environmental impacts, such as carbon dioxide and nitrous oxide emissions when driving a car or using combustibles for heating or cooking. By far the most environmental impacts of consumption, however, are indirect. This includes using electricity for household appliances, which does, in itself, not produce emissions or waste, but the impacts depend on the source of electricity used. Even more indirect are the impacts further down the production chain, for instance the emissions and amounts of waste embedded in a computer someone buys, during the production of which raw materials have been extracted and energy has been used[158].
To sum up, my use of the term consumption here broadly follows Stern's working definition, with the additional remark that I use two different perspectives on consumption: the macroeconomic perspective and the one on final consumption by individuals or households. The latter includes both acts of purchasing and use patterns, and it takes into account both direct and indirect (embedded) impacts.

156 Cf. the statement by Princen et al. (2002: 16): "Producers are consumers; production is consumption".

157 As Schramme (2011: 345) has remarked, the "lack of consumption" of goods that have been produced can be seen as a form of misconsumption. I do not distinguish here between goods that have been produced and are actually consumed and those that go to waste without human use (e.g. expired food), because I consider the latter to be driven by existing consumption patterns, too, and thus closely linked to the former. Also, I do not know how relevant this issue is from a quantitative perspective, as, except for the example of food or other items with a short lifespan, probably most consumption goods are offered to consumers over time at ever cheaper prices until someone buys and thus "consumes" them in the end.

158 It is worth mentioning in this context that imports and exports of goods and services inhibit equating the impacts of national production with those of national consumption (EEA 2013 a: 32). Thus, the adoption of a consumption perspective makes it more difficult to determine national shares of globally relevant emissions, such as carbon dioxide.

Social innovations

A final concept that is frequently mentioned in the context of sustainable consumption and that is relevant for the subject here is that of "social innovations" (cf. e.g. Zapf 1989, Howaldt/Schwarz 2010: 87 ff., Rückert-John et al. 2013, Deutscher Bundestag 2013: 517, UBA 2014 a). An innovation is commonly understood as something new—most often a process, technical artefact, or an organisational change—that helps someone to reach a certain target more easily or effectively (cf. Zapf 1989: 173 f.). As the term "innovation" is most often applied to technical changes, it seems that the recent increase in publications on explicitly *social* innovations within the sustainability debate is linked to the insight that technical solutions are not enough to achieve the aims of sustainability (cf. Rückert-John et al. 2013: 5, Deutscher Bundestag 2013: 517). Social innovations are thus, like sufficiency, compared to and conceptually separated from technical innovations (cf. Howaldt/Schwarz 2010: 7 ff.). As Zapf notes, social innovations can be preconditions, concomitants, or consequences of technical innovations; while technical innovations are based on the manipulation of energy and materials (Zapf 1989: 177), social innovations rather involve the development or mainstreaming of new social practices. This may include new forms of organisation, new regulations, or new lifestyles, and comprises not only top-down reforms but also everyday practices (Zapf 1989: 177). The transition between technical and social innovations seems to be rather smooth (cf. also section 3.1.5), as examples from the discussion on sustainable consumption show: the success of many social innovations in the area of consumption relies on technical possibilities, such as organisation via the internet or smartphones. Examples of social innovations in the field of consumption include couchsurfing, carsharing, clothing exchange platforms, repair cafés, urban gardening, and websites that disclose places where unowned food can be harvested (cf. Rückert-John et al. 2013: 24 ff., UBA 2014 a). As many social innovations are based on the practice of sharing goods, some see them as signs of a new "sharing" or "collaborative" economy (e.g. Heinrichs 2013 b, cf. also EC 2016).

The distinction between social and technical innovations reminds us, of course, of the preceding discussion on technology and sufficiency; therefore, the question arises of what the difference between social innovations and sufficiency is or how they are connected. As social innovations explicitly deal with novelties[159], sufficiency seems to be broader as it includes not only new but all practices that lead to reduced consumption with high environmental impacts. Also, the "social" aspect in social innovations indicates that these are rather not individualistic but group-based practices, while sufficiency can often be realised by individuals with regard to their own lifestyle (e.g. saving heating energy at home, not eating meat). The following statement gives a helpful indication of the relationship between the two concepts: "Social innovations can make sure that sufficiency strategies can be brought in line with adequate welfare and a high quality of life" (Deutscher Bundestag 2013: 517)[160]. This means that sufficiency aims at limiting individual consumption of certain goods and services, while social innovations still allow for access to these when they are really needed. An example in this sense is, again, carsharing: if someone usually does without a car, through the social innovation of carsharing this person has access to a car in situations where it really makes life easier.

3.3 Interim summary

This chapter has presented and analysed sufficiency and technical solutions as sustainability strategies. As regards terminology, both concepts are not easy to define, and in particular the term sufficiency is used in numerous interpretations in academic literature. Thus, unlike other authors have attempted to do, I do not propose a clear demarcation of sufficiency, but rather a broad definition that allows for a rough distinction between it and technical solutions: *Sufficiency is the reduction of demand or use of goods and services with high environmental impacts in order to achieve per-capita consumption levels that ensure emissions and resource use to stay within the environmental carrying capacity.* This vague definition does not exclude all overlaps with technical strategies; however, it still

159 Actually, in the discussion on social innovations very often practices are mentioned under this header that are as old as civilisation: sharing goods, reselling goods that are not needed anymore, collecting food that others intend to waste etc. (cf. Rückert-John et al. 2013). The novelty often lies in the degree and form of organisation, as well as in the fact that people do this out of conviction rather than poverty.

160 My translation.

seems to be appropriate for the following analysis, as the intention is not to play off one strategy against the other, but to work towards the common aim of reducing absolute environmental impacts. In addition to this definition, juxtaposing sufficiency and technical solutions in relation to several other concepts, as done in Table 4, further helps us to understand the differences. While sufficiency focuses more, though not exclusively, on consumption, demand, subjects, behaviour change, and absolute limits, technical solutions focus more on production, supply, objects, technical change, and relative improvements.

For further clarification, a number of concepts that are similar or have at least been linked to sufficiency in the literature on this subject have been presented, including the concepts of sufficiency as a notion in justice theory, satisficing, sustainable consumption, and social innovations. They were all compared to and distinguished from sufficiency (as I use it here), with the result that they all feature interesting links to or perspectives on the sufficiency debate, but none of them means exactly the same as sufficiency.

This chapter has also shown that many environmentalists, at least in the past, often tended to fiercely advocate one of the strategies, while denying the problem-solving capacity of the other. Lately, however, the voices that suggest that both are needed to curb the environmental impacts of consumption patterns in wealthy countries to sustainable levels have become louder. The preceding analysis has defended the position that both strategies are needed, as both have their limitations: for technology, these include in particular the fact that no technology can be extended in its application infinitely, as there are conflicts with other societal aims as well as limits, for example in the form of suitable space; that all technologies have side effects, including environmental impacts, which also limits public acceptance of their application; that there are environmental problems that cannot be solved by means of technology ("commons" problems); and that the development and diffusion of respective technologies is expensive and requires time. Further, the effectiveness of both strategies is hampered by rebound effects, although to a lesser degree in the case of sufficiency. A major limitation to sufficiency are problems of acceptance, because it can have more significant impacts on everyday habits and because there is uncertainty about its effects on the economy and the labour market if it is adopted by a relevant number of people. Finally, an additional weakness of sufficiency is that considerable time would pass for the strategy to reveal its relevant effects until the value change required has happened if the strategy is supposed to be based mainly on individual, voluntary adoption of sufficiency practices.

It has been mentioned occasionally that mainly the technical strategy has been implemented so far by means of policies, and in particular technical solutions (besides organisational aspects) are the focus of political programmes such as "Ecological Modernisation", the "Green New Deal", "Green Economy", or "Third Industrial Revolution". The next chapter will therefore concentrate on sufficiency and its role as a strategy of individuals as well as its potential as a policy.

4. The individual and the political

> *"'Environmental protection starts at home!' This is supposed to say: whatever pollution and destruction there is to bemoan, we are all to blame, as we are the ones who drive cars and use sprays, wash the laundry and carry home plastic bags, demand hygienic packaging and drink juice from aluminium cans. [...] We cringe, conscious of our guilt, and find all this sounds quite plausible—however, it is sheer cynicism, it derides reality as it is, and therefore derides us, who are obviously believed not to be able to realise that the intention is just to distract us. [...] The appeal that environmental protection starts at home corresponds to the advice to a handcuffed person that he could, after all, still blink with his eyelids. In other words: There are only a few very limited areas of life, within which the individual can contribute to so-called environmental protection. Beyond these, modern life masters us to an extent that we may not hope at all that we could put that slogan into practice in a way which changes the world and saves the environment. That environmental protection starts at home derides the generally accepted and effective norms of production, of consumption, of life at all, to which we are widely bound in obediance for good or for evil [...]."* (Dahl 1991: 229 f.)[161]

Introduction

Like Jürgen Dahl in the above quotation taken from his essay *Zwölf Zylinder, schadstoffarm*[162] (1991), many authors have expressed concerns about the practice of focusing on individual behaviour in environmental protection. While Dahl was certainly too pessimistic, as the following chapter will show, he still points to an important issue: that individual efforts supposed to contribute to environmental protection cannot replace collective efforts.

The preceding chapter pointed out why sufficiency is necessary for achieving sustainability. Questions that typically arise in the context of the sufficiency discussion are often located in a "grey area" between the regulation of social interactions and individual decisions for one's own life (Ott/Voget 2007). In the past, most authors advocating the sufficiency

---

161 My translation.

162 "Twelve cylinders, low-polluting".

strategy focused their analysis on the individual[163]. Typical examples of individual sufficiency practices are: not using a car but public transport or a bicycle, eating less meat, or abstaining from holidays in exotic countries that can be reached only by aeroplane and spending the time in a nearby place instead. These kinds of individual measures, however, might be not effective enough to bring about the required reduction in environmental impacts.

For technical strategies to protect the environment, political interventions encouraging technological development or even making a certain technical minimum standard mandatory are, by now, implemented and commonly accepted in many countries, even though the level of ambition of these policies is often subject to tough discussions. For instance, in the European Union regulations are in place on the air emissions allowed from industrial plants, on the post-consumer treatment of different sorts of waste, and on the minimum energy efficiency of many electric devices. However, it is much more difficult politically to call into question the necessity of kilometres driven by car or the amounts of electric devices that are being demanded.[164] A discussion on the necessity of implementing the sufficiency strategy on the political level has only gained momentum in recent years.[165]

This section summarises the benefits of individual practices in the sense of sufficiency as well as their practical limitations in order to show why individual approaches might not be enough. It subsequently gives an overview of the discussion in the literature on sufficiency as policy as well as the related concerns, before providing my own in-depth analysis in the following sections (from section 4.2 onwards).

---

163 See, for instance, Linz et al. (2002, with the exception of Winterfeld's contribution), Kleinhückelkotten (2005), and Stengel (2011).

164 It should, however, be mentioned that some policies, such as ecological taxes on electricity, encourage not only higher energy efficiency, but also generally influence demand in the sense of sufficiency (cf. chapter 5). There is, however, no systematic political implementation of sufficiency, as opposed to the technology strategy.

165 Rather early publications include Winterfeld (2002) and Linz (2004: 36 ff.). For more recent ones, see Darby (2007), Linz/Kristof (2007), Ott/Döring (2007), Ott/Voget (2007), Müller (2009), Voget (2009), Brischke/Spengler (2011), Linz/Scherhorn (2011), Linz (2012 a, 2015, 2017), Seidl/Zahrnt (2012), Voget-Kleschin (2012: 106 ff.), Heyen et al. (2013), Spengler (2013), and Schneidewind/Zahrnt (2014); cf. also section 4.1.2.

Before entering the analysis of benefits and limitations of individual and political approaches to sufficiency, it needs to be clarified what is meant here by the terms "individual" and "political".

"Individual" and "political"

"Individual", as I use it here, can refer to single persons, but also to single households and to private initiatives encapsulating a rather small number of people (e.g. in neighbourhoods). This is different from the use of "individual" in liberalism, where it is always referred to individual persons, who are the central entities of concern for liberals[166]. Since I use liberalism as the framework for my discussion, this different use of the term might be a problem; however, I think this is not the case as the point here is not to assign, for example, rights to households (besides individuals), but to discuss the effectiveness of sustainable behaviour of single entities, such as individuals, households or other small private groups, versus the effectiveness of (large-scale) collective action for sustainability. For this purpose, it does not matter much whether the single entities are individuals or small groups of people.

What does "political" mean in this context? There are many different aspects of this term that can be highlighted in order to distinguish the political from the non-political. For instance, there are well-known approaches to the term, such as Max Weber's interpretation related to the distribution of political power (Weber 1992 [1919]) or Carl Schmitt's explanation of political activities by means of his friend-enemy distinction (Schmitt 1991 [1932]). As I use it here, what is crucial in the notion of "political" is that political action aims at the *collective* level[167]. In this sense, Young has used the following understanding of the term:

> *"[...] by politics or the political I am referring to the activity in which people organize collectively to regulate or transform some aspect of their shared social conditions, along with the communicative activities in which they try to*

166 This is above all true for traditional views of liberalism (cf. Talisse 2005: 16 ff.). However, it should be noted that there are also comprehensive discussions on the role of communities in liberalism (e.g. Raz 1986: 17 f., Miller 1999 b: 174).

167 Cf. also common translations of the ancient Greek word *ta politika* as "the public, the common, what concerns all citizens" (Schmidt 2004: 538).

persuade one another to join such collective action or decide what direction they wish to take it." (Young 2004: 377)

As Young puts it, this is "broader than government" (2004: 377). My use of the political within this study shares Young's focus on collective organisation, but it does not share the idea of going beyond the activities of the state on its various administrative levels, including its commitments made on the international level. This is not because I consider non-state "political" activities as unimportant, quite the contrary. The reason for this limitation is that activities by, for instance, private organisations which are directed at the policy level would require quite different reasoning in terms of their justification than what I propose here for the state (cf. section 4.2). Moreover, this limitation simply corresponds to the focus of this study, which lies on collectively binding decisions instead of considering all activities which are directed at the policy level[168]. This also means that my use of "political" goes beyond what is voluntarily applied and thus includes measures that are potentially coercive.

Such decisions or agreements, as political scientist Elinor Ostrom has pointed out, can also apply collectively within groups which have managed to organise themselves without government interference (Ostrom 1990). However, for the field of consumption it is unlikely that such private agreements would work because this is not a field concerned with a geographically limited resource, such as the famous pasture, and it is not just a limited number of parties that is involved. I therefore also exclude from my analysis collectively binding but privately taken measures.[169]

168 Such as activities by individuals, for example, as expressed in the notion of "political consumption" (cf. Heidbrink/Schmidt 2011: 32 ff.).

169 I use mainly "individual" versus "political" in the following, but there are also other pairs of terms that are used in a similar sense in the discussion, including private/public, individual/collective, voluntary/coercive. These will also be used where appropriate in the following analysis.

4.1 Sufficiency as an individual and as a political approach

4.1.1 Sufficiency on individual level: practical potential and limitations

Potential

Individual sufficiency practices are important from a sustainability perspective and can have benefits for the individual as well as the society as a whole.

First, individual action can go beyond what is required and can therefore realise an added value. From a quantitative perspective, private consumption is directly responsible for a large share of the environmental load, and an even much larger share if indirect effects resulting from the production chain are included. For instance, with regard to greenhouse gas emissions, the direct share of private consumption is 40% (Reusswig/Battaglini 2006: 6). Thus, individuals can make a difference, for instance, with regard to the potential for energy saving through lifestyle changes (Siebenhüner 2011). Households can reduce their electricity consumption by up to 80% compared to the average if energy efficient appliances are used and "energy sufficient" practices are applied (Brischke et al. 2011).

Second, sufficiency practices can have positive effects on individuals in terms of a healthier or happier life if a person decides to change his or her life in favour of "slowing down" strategies, such as less working time, which entails less income but more leisure time. Decreasing one's amount of possessions and curbing one's endeavours to acquire new ones can have disburdening effects on individuals (cf. discussion in Jackson 2005). Such alternative views of the good life have been promoted since ancient Greece and, more recently, by the "Voluntary Simplicity" movement (Voget-Kleschin 2012: 76 ff., cf. section 3.2.1) and in the form of "Alternative Hedonism" (Soper 2007, 2008 a, 2008 b). In addition, if groups of individuals implement sufficiency measures on a local scale, this can locally reduce environmental effects caused, for example, by harmful emissions or waste (Alcott 2008: 782, Princen 2005).

Third, individual behaviour change can encourage others to follow the example they are given (Ott/Voget 2007). It can demonstrate what is actually possible and dispel doubts, for example, about whether it is at all feasible to live without a car as a family with three children. Such individuals and pioneers from civil society can play a "pacesetter role" in social change (Linz 2004: 34).

In line with this and fourth, individual action can send signals to markets and decision-makers, in particular when they form a trend ("voting with dollars") or become visible in statistical indicators that may be interpreted as a level of acceptance for political interventions (Bilharz et al. 2011). It must, however, be noted that consumption and "non-consumption" acts differ with regard to their potential to send such signals. Non-consumption actis can be considered to be less visible than, for instance, buying eco-labelled or energy-efficient products. Moreover, the motivation is less obvious: someone can avoid using a car out of conviction about its damaging impact or because he or she simply cannot afford a car and would in fact like to have one.

One could subsume the more *political* activities of individuals under this point, too. A popular categorisation distinguishes the "individual as citizen" as opposed to the "individual as consumer" (cf. e.g. Berglund/Matti 2006, Dobson 2010, Maniates 2002, Reusswig/Battaglini 2006, Sagoff 1988: 7 f.). Reusswig and Battaglini describe the former as being driven by reflective and policy preferences, whereas the latter expresses market preferences (2006: 11 ff.)[170]. Examples of typical "individual as citizen"-like activities are participation in signature initiatives, in civil society groups which organise consumer information campaigns, in consumer protests against companies that act in a socially or environmentally unsound way, or in "critical mass" movements that demand better conditions for cyclists in urban traffic infrastructures. Many consumption researchers have argued that neither of these two faces of individuals should be neglected.[171]

However, most individual "citizen" activities need to be distinguished from individual acts that directly aim to reduce adverse environmental impacts. This is because they do not aim to reduce impacts in the first place, but change the conditions on the collective level. If successful, in a second

---

170 Cf. also the characterisation by Berglund and Matti (2006: 555), which has a much stronger tendency to label the two as good vs. bad: in their account, citizens are altruistic and community-oriented, while consumers are materialistic and focus on individual short-term gains.

171 As opposed to this, Soper states that the split between "consumer" and "citizen" is not adequate, as it cannot explain actions that could be attributed to both fields, such as product boycotts (Soper 2007). She also argues that a perspective in which "the consumer is conceived as the cause of the unfortunate environmental consequences that the citizen as active campaigner will attempt to correct or alleviate" (2007: 208) does not reflect individuals adequately.

step, according collective measures will be taken. Thus, "citizen" action often aims at political changes because individual "consumer" action is not effective enough.

Limitations

There are also a number of weaknesses that limit the effectiveness of individual sufficiency practices and hinder them from becoming more accepted[172]. This first includes the frustration that may take hold if the individual notes that his or her "ecologically correct" behaviour has very limited effects as a result of the "business as usual" attitude of the majority (Maniates 2002: 58 f.; see also Sinnott-Armstrong 2010: 344). If only 10% of the members of a society are disposed to limit their resource intensive consumption activities, this is simply not enough from a quantitative perspective (Ott/Voget 2007). In addition, some things simply cannot be influenced by individuals: for instance, around 10% of per-capita carbon dioxide emissions in Germany are caused by the construction of public infrastructure (UBA 2014 b: 10).

Second, environmental processes are complex and so are the often long and unclear production chains of products. To complicate things further, the approaches and mechanisms societies have implemented to deal with adverse environmental impacts are often difficult to understand as well and can affect the usefulness of well-meant individual actions (Grunwald 2010: 179). Therefore, the knowledge of individuals about the consequences of their behaviour must remain rather limited. Attaching information and environmental labels to products can only provide some relief in this regard (Petersen/Schiller 2011, Grunwald 2010). And even if individuals are aware of the ecological consequences of their consumption acts, they often do not change their behaviour accordingly. This "knowledge–behaviour gap" is well-known in environmental sociology and psychology (e.g. Kollmuss/Agyeman 2002).

Third, macroeconomic rebound effects, which can, in principle, result from all kinds of micro-level energy and resource saving measures, may limit the effectiveness of individual sufficiency measures because of inter-

172 Also, cf. the general limitations to sufficiency (section 3.2.4).

mediary levels that absorb the intended effect (Alcott 2008, Grunwald 2010, cf. section 3.2.4).

Fourth, individual action can be used as excuse not to take collective action. Political theorist Murray Bookchin already diagnosed the "privatisation of the environmental crisis" in 1989, stating: "If 'simple living' and militant recycling are the main solutions to the environmental casts, the crisis will certainly continue and intensify" (Bookchin 1989). In the same vein, Maniates observes some "individualisation" of the task of solving environmental problems ("Plant a tree, buy a bike, save the world?"), while at the same time other necessary responses, including the development of according institutions, are disregarded (Maniates 2002: 45 f.). Even more critically, Winterfeld notes that without calling into question the predominant economic model, private sufficiency action could rather *support* and help to maintain the growth economy (Winterfeld 2002: 34). In this sense, individual sufficiency practices can be interpreted as having "unusual utility functions" for the rest of society (cf. Ott/Voget 2007: 27), as individual renunciation of material wealth enables others to maintain or increase their own.

Fifth, the given socio-economic and legal conditions, or "systems of provision" (Spaargaren/van Vliet 2000, Spaargaren 2009), determine the latitude of individual action and influence the decisions taken by individuals. This area has been subject to comprehensive analyses by consumption researchers[173]. Numerous proposals exist that try to categorise the different forces that drive consumption (cf. Shove/Warde 1998, Røpke 1999: 404 ff., Sanne 2002: 276 ff., EEA 2012 b: 20 ff.). Referencing Veblen's famous notion of "conspicuous consumption" (Veblen 2007 [1899]), research in this field has long concentrated on psychological and social drivers related to visible consumption that conveys status and meaning (Shove/Warde 1998). Shove and Warde (1998: 11 ff.) note, however, that an important part of environmentally significant consumption cannot be explained via such drivers, but is rather related to "inconspicuous consumption" practices brought about by the infrastructure given, effects of interdependence, and "the creeping evolution of normal standards". Some of these drivers are steered by political decisions directly, others can only be influenced (if at all) indirectly.

173 See e.g. Shove/Warde (1998), Røpke (1999), Sanne (2002), Schor (2005 a); see also Winterfeld (1993), Spaargaren/van Vliet (2000: 65), Maniates (2002: 65), Spaargaren (2009), Voget (2009: 218 ff.), EEA (2012 b: 20 ff.).

The most obvious example are transport infrastructures, which are defined in political planning processes. The availability and comfort of public transport, bicycle lanes, the number of parking lots, the likelihood of congestion, as well as their associated costs have an important influence on people's choices. Besides the transport infrastructures themselves, other planning decisions, such as defining areas of low-density one-familiy housing, can, besides other environmentally damaging effects, result in additional traffic (Sanne 2002: 277).

In addition, there are various economic drivers of consumption. Due to a lack of internalisation of external costs, the prices of resource-intensive goods are relatively low compared to labour-intensive goods or services, which creates an incentive to consume more of the former rather than the latter (Røpke 1999: 405, EEA 2012 b: 20). This is the case to an even stronger degree for the prices of imported commodities due to lower environmental standards *and* low labour costs in the countries of origin (Schor 2005 a). Sometimes the existing system of taxes and subsidies can also encourage individuals to consume more (EEA 2012 b: 21). Such incentives include, for instance, taxation rules that allow to offset costs for goods or commuter travelling costs against tax.

Some other drivers for individuals to consume more are related to the practices of businesses that have an interest in selling more products, which are, again, embedded in the structures given and the pressures of the market. The main pressure on companies in this regard is competition, which leads to advertising and techniques to make offers more attractive (e.g. credits given to consumers by retailers), as well as to speeding up innovation cycles together with faster consumption cycles (Røpke 1999: 404). Further, the structure of the labour market in many countries encourages companies to increase wages as a result of productivity gains instead of reducing the working hours of their employees. Workers then take the additional income and spend it on consumption (Røpke 1999: 404 f., Sanne 2002: 279 f., Schor 2005 b).

Finally, at an accelerated pace of technical development, technical systems change more frequently. Some consumption items come in clusters, so that when the system changes, all items need to be exchanged and old components that still function cannot be used any longer because interfaces between devices have changed (Sanne 2002: 278).

Sanne (2002: 273 f.) argues that consumers are, on the one hand, not helpless "victims" who can only follow the marketing strategies of producers; however, it should also not be neglected that companies often pressure governments to "create conducive conditions for increasing consumption" and that they make intentional use of the drivers which make people consume more. Thus, individuals can in principle resist incentives to consume more to some degree, but they must be rather strong-willed to do so.

4.1.2 Sufficiency as a political concern: state of the debate

Because of the given limitations, literature dealing with sufficiency increasingly often raises the point that individual strategies are not enough and sufficiency should also become a political concern.[174] However, so far such statements have been backed mainly with the authors' (anecdotal) observations and not with a systematic discussion of the rationale and implications a sufficiency policy would have.[175] This section summarises the current state of the debate.[176] My own approach will be presented from section 4.2 onwards.

Potential role and functions of a sufficiency policy

The role which sufficiency advocates assign to policy is the provision of "corrected" framework conditions in order to encourage, support, or "nudge"[177] individual decision-making towards sufficiency. Less often, regulatory approaches that require consumers to act more in line with sufficiency are also called for. Accordingly, the proposals range from enabling strategies to more behaviour-determining approaches such as economic signals and regulatory limits (cf. also section 5.1.1).

General statements that a broader change in values and acceptance of sufficiency require an appropriate political and social framework can frequently be found (e.g. Hennicke 2002: 58, Linz 2004, Kleinhückelkotten

174 Cf. footnote 165.

175 With the exception of Voget-Kleschin (2012: 106 ff.), who offers a somewhat more comprehensive discussion.

176 Note that this section deals only with publications explicitly referring to the concept of sufficiency in the sense used here. Other publications that discuss the general issue of the options and legitimacy of restricting consumers' choices will be taken into account at a later stage.

177 Cf. Thaler and Sunstein (2008); see also section 5.1.

2005: 64 f., Scherhorn 2008: 3, Voget 2009: 221). Most authors state that lifestyles cannot be regulated directly in a liberal society, but that influencing individual consumer behaviour is possible or even required (cf. e.g. Ott/Voget 2007, Voget-Kleschin 2012: 96).

Winterfeld's "Right to Sufficiency", which demands that "No one should have to always want more" (Winterfeld 2002: 31),[178] is located at the end of enabling strategies. She states that not wanting something, for instance a new computer, given the fast innovation cycles, requires an immense effort by the individual. The function of policy would therefore be to facilitate and protect sufficiency. Similarly, Seidl and Zahrnt (2012) and Schneidewind and Zahrnt (2014) call for a policy that provides the conditions which make it easier to realise the "simple" life, so that environmentally sound consumption patterns and lifestyles can evolve.

A step further, a number of authors hold that privileging behaviour which is in line with sufficiency (Ott/Voget 2007: 20) and even steering consumption towards sufficiency through economic signals (including taxation, subsidies) would be possible in a liberal society. Huber (1995 b: 40) suggests internalising external environmental costs to steer demand to be in line with sufficiency.

Besides setting incentives, some authors state that the policy level could also use regulatory law and planning instruments in order to achieve sufficiency (e.g. Linz 2004: 36, Linz/Kristof 2007: 187 f., Linz 2012). Linz (2012) highlights two instruments for a sufficiency policy. Besides steering via prices, this includes setting limits, such as impact caps and speed limits.

Besides these, a number of authors propose concrete policy instruments in order to realise sufficiency as policy. I will come back to these in chapter 5. Some authors think that the objections to sufficiency as policy are too strong and thus conclude that political implementation of the sufficiency strategy would only be likely when the environmental conditions have become so much worse that any adverse effects of sufficiency would be chosen as the lesser evil (Huber 1995 a: 129, Stengel 2011: 235). The next two sections will deal with such objections, which have been raised with regard to the potential impact of a sufficiency policy on individual liberty as well as in relation to its practical feasibility and potential consequences.

178 My translation.

Concerns about the impact of a sufficiency policy on liberty and according responses

As regards the concerns about liberty, most writers recognise that this is a critical issue which must be examined carefully in order to justify interference with individual freedom and dispel concerns that sufficiency policy could lead to an "eco-dictatorship" (e.g. Huber 2000). Some have also developed responses to these concerns; the discussion remains, however, mostly rather sketchy.

Mainly two forms of reasoning are discussed in the literature on this subject: on the one hand, specifying an account of the good life and designing according policies (cf. section 2.1.3), and on the other, referring to the notion of harm (cf. section 4.2.1). With regard to the first one, Winterfeld (2002) sees a dilemma in political approaches to sufficiency. She argues that, on the one hand, such approaches are necessary, as individual-level sufficiency is not effective enough. On the other hand, she notes that prescribing sufficiency measures on the political level and, in the sense defined by her, thus determining some clear picture of the good life is not possible (cf. also Müller 2009), as separating individual morality and policy is an important achievement of modernity. Diluting this separation would imply "opening the floodgates to political arbitrariness", since "in the name of morality, many people have been burned, beheaded, stoned to death or otherwise executed" (Winterfeld 2002: 34 f.)[179]. That modern states leave the decision about the "morally correct" way of life to individuals[180] is, according to Winterfeld, a positive development from this perspective. As a way out of the dilemma, Winterfeld refers to Nussbaum's capabilities approach (cf. Nussbaum 2011), suggesting that sufficiency should be considered a human capability, which has to be facilitated and

179 My translation.

180 Note that the kind of morality Winterfeld refers to suggests, however, that she is mixing different types of actions which would deserve clearer separation corresponding to their consequences. The examples of political arbitrariness she has in mind are obviously rather related to the most personal freedoms, for instance religion or sexual orientation, which are today accepted as being of a strictly personal nature in liberal societies, as they do not have negative impacts on others. The kind of behaviour producing adverse environmental impacts, however, would belong to a different category (see section 4.2). Nevertheless, Winterfeld raises an important point here: the moralisation of individual, allegedly environmentally "wrong" actions, which will be discussed in section 4.1.5.

protected by the state. Other authors, such as Schneidewind and Zahrnt (2014), think that sufficiency policy can be based on some image of the good, and that the good life actually is a political issue. The emphasis of their account lies in the idea that policies should create "the conditions in which the greatest number of individual life concepts can be realised" (Schneidewind/Zahrnt 2014: 12). What remains unclear, however, is why these individual life concepts should necessarily be in line with sustainability, and what should be done if not.

With regard to the second way of arguing based on the notion of harm, there are several authors who mention this as a possible justification for sufficiency policy. Müller (2009), for instance, considers this approach to be feasible, as the duty of not harming others is a commonly accepted minimal baseline within liberalism. This principle alone is, in Müller's view, enough to argue for the necessity of a sufficiency policy in liberal societies, given current assessments of the likely impacts of climate change in the future and the insufficient effectiveness of technical solutions (2009: 87 f.). He concludes that sufficiency "is similarly adequate for a liberal society as efficiency" (2009: 88). In a similar way, Stengel (2011: 326 ff.) draws on the duty of individuals not to prevent others from exercising their freedoms and rights. Based on this and some policy examples, he concludes that the liberal state does not have to wait for voluntary implementation of the desired behaviour, but may or even must define rules for the protection of the population (Stengel 2011: 328 f.). However, somewhat more sceptically, he also notes that some consumption acts have complex chains of impacts, which may be very distant from a consumption act itself in space and time and are problematic only as the sum of many similar acts, which complicates things further. Finally, Voget-Kleschin also discusses the harm principle as a basis from which to restrict consumers' choices and concludes that it is legitimate to do so (2012: 106 ff., 156). In addition, as an argument against a simple refusal of government interference in consumers' choices, she mentions that these choices are affected by the structural surroundings which are provided by the society and the state anyway. She states that therefore, "objecting to any and all governmental interference with people's choices is (logically) impossible" (Voget-Kleschin 2012: 107). While this is an argument supporting the view that the effectiveness of individual actions alone is limited (cf. section 4.1.1), it cannot be used as a stand-alone justification for a sufficiency policy. The fact that the state does influence consumer decisions anyway provides no justification in itself of why it should influence them *in a certain direction.* Otherwise one could also suggest that it would be legitimate if the state, say, tried to make consumers buy only domestically produced goods.

Both options with which to justify a sufficiency policy—harm-based and good life approaches—, as well as other ones, will be discussed further in later chapters (cf. sections 4.2 and 4.4).

Practical concerns

In terms of practical concerns there are to mention, first, doubts about the feasibility and social acceptance of a sufficiency policy within a democracy, second, uncertainty about whether this is at all a field which can be steered via policies and, third, concerns with regard to the potential economic and social consequences if such a policy was realised.

As regards feasibility and acceptance, political decisions obviously need to be backed by acceptance among the population in democracies. Measures that have beneficial effects only in the (very) long term are often rejected due to options available which unfold their benefits closer in time (Stengel 2011: 228). Thus, as Stengel states, if politicians dared to promote long-term preventive measures, they would quickly lose votes (cf. also Heyen et al. 2013: 18). He concludes that the sufficiency strategy would receive political support only if it either did not jeopardise economic growth (which he considers unlikely) or if the environmental situation became so poor that the alternative of a lack of economic growth would be considered the lesser evil (which he considers likely) (Stengel 2011: 235). Stengel therefore suggests civil society as the main driver of sufficiency rather than the state or other actors (2011: 344). In a similar way, Huber states that the sufficiency strategy's "potential of resonance and compatibility" within society is far too low to base a policy on it. As long as the factual circumstances of a crisis are not compelling enough to adopt sufficiency practices, the strategy, according to him, remains unrealistic. To act as this compelling force through policy would be beyond any understanding of democracy. Huber therefore concludes that, from a political perspective, spending too much time on the idea of sufficiency is dispensable (Huber 1995 a: 129 f., 1999: 12).

In a more positive manner, Linz and Kristof (2007: 188) acknowledge the problems of sufficiency as policy in terms of its acceptance, but in consequence, instead of rejecting the overall possibility, they outline ways to overcome the (initial) resistance from those not willing to accept the changes related to sufficiency. They suggest that a discussion is needed, public pressure has to be formed, and that parliaments and parties need to be involved, so that in the end a majority supports the required changes or at least does not oppose them (Linz/Kristof 2007: 188).

Besides problems related to acceptance, there are doubts about whether something such as sufficiency can be brought about at all by political steering[181], given the complexity of the socio-technical systems addressed and existing material and cognitive path dependencies (Heyen et al. 2013: 17). First, there are certain general problems related to political steering: steering efforts can be affected by implementation deficits, the unwillingness of those concerned to accept them, or the various interests present within a multilevel administrative system. Second, besides these, steering with regard to environmental issues poses further challenges (SRU 2012: 369). These include the difficulties of dealing with common resources, particularly in the case of global issues, the often existing gap in time between the causation and the effects of environmental problems, and the limited degree to which individuals can actually perceive current environmental problems in industrialised nations (as opposed to e.g. air and water pollution, and also because production has been displaced to other countries). On the other hand, as has been mentioned before, individual behaviour is embedded within systems that are shaped, among other things, by political decisions (cf. section 4.1.1). To use the example of infrastructures again, it is clear that policies do have an influence on the choice of means of transport in a city (cf. section 5.2.3). In such cases, political steering should in principle be effective, depending on the selection and concrete design of a policy. In any case, there are important lessons from research on political steering, including that it is necessary to look at the bigger picture and anticipate potential side effects, such as the mere replacement of impacts (including rebound effects, cf. section 3.2.4), or other undesirable consequences of a policy.

Finally, it is unknown what the economic and, in consequence, also the social impacts of society-wide implementation of the sufficiency strategy

181 Political steering describes the attempt by political actors to design or influence social systems in terms of certain targets or purposes. As social scientists in the US and Europe have found in empirical studies since the 1960 s, acts of political steering do not necessarily imply success in reaching their targets (Mayntz 1987, 2004).

would be (cf. Linz 2006). Quite in contrast to the demands of sufficiency advocates, increasing consumption is often seen as a means of overcoming social and economic problems of whatever kind, which is also reflected in the revival of "Keynesian" political economics after the economic downturn in 2008 (Stengel 2011: 161). A comprehensive analysis of the potential economic consequences of the sufficiency strategy is lacking so far. However, it is obvious that a *sudden* turn to a widely applied, national sufficiency policy would have widespread impacts on a country's economy, very likely resulting in a grave recession. And it is not only the national economy that might be affected negatively; global linkages, such as the dependency of developing countries on exports to wealthy countries, also have to be taken into account (Stengel 2011: 163). Linz (2006) asks some relevant questions in terms of the potential economic consequences of sufficiency: what will happen to the economy? Does sufficiency lead to a lower overall economic volume? Does it cost jobs? His conclusion that a development in the sense of sufficiency does not have to harm the economy (Linz 2006: 21) is not (yet) backed by sufficient scientific evidence.

Nevertheless, there is a growing body of literature on the economic and social implications of a potential "beyond growth" or "de-growth" economy (e.g. Rudolph 2007, Jackson 2009, Seidl/Zahrnt 2010; cf. section 2.3.5). Some lessons can be derived from this debate for the question regarding the implications of sufficiency. However, the requests put forward by "beyond growth" advocates and the requests related to the idea of sufficiency are not exactly the same, and therefore the potential consequences are also different: while criticism of economic growth requests shifting the focus from GDP growth as a political goal to other things (cf. section 2.3.5), sufficiency advocates request a decrease in environmentally harmful consumption. Still, even though the emphasis of these two debates is different, implementing sufficiency as policy could also lead to various paths in terms of GDP, including the possibility of maintaining the same level of GDP by restructuring the economy and shifting the focus from resource-intensive consumption to the service sector. In practice, however, this would be a rather long and, for some, painful process.

The concerns mentioned need to be taken seriously. While this study concentrates on a discussion of the concerns with regard to the justification needed (sections 4.2 to 4.4) and the question of acceptance in a democracy will also be dealt with briefly (section 4.4.4), a comprehensive discussion of the success factors for political steering in the sense of sufficiency as well as of its potential economic consequences is beyond my scope here.

4.1.3 Interdependence between the individual and collective levels

In the context of the discussion of the role of the social structures or "systems of provision" for individual consumer behaviour, consumption researchers have also pointed out that not only do the structures influence the individual, but, conversely, individuals also influence and shape these structures through their adoption and continued performance of certain practices (see e.g. Spaargaren 2000, Spaargaren/Vliet 2000, Reusswig/Battaglini 2006: 11 ff., Bilharz/Fricke 2011, Bilharz et al. 2011). This idea draws on Anthony Giddens' "Structuration Theory" (Giddens 1984), in which he proclaimed the "duality of structure"—the theorem that agents and structures do not exist independently of each other but represent a "duality". This means that structures do not only constrain and enable individual action, but that individuals also reproduce the structural properties of societies through their daily routines. "Structure", in Giddens' sense, does not only refer to external laws or infrastructures, but also to other "rules", including normative elements and codes of signification, and "resources", including authoritative and allocative ones directed at the control of human agents and "aspects of the material world" (Giddens 1984: xxxi).

Individual routines are, for instance, shaped by social standards of "comfort, cleanliness and convenience" (Shove 2003), which are common standards learned and shared with many others and reproduced in the course of their adoption and performance (Spaargaren/Vliet 2000: 69). Individual action thus consolidates existing structures, but it can also challenge and change them if a "critical mass" of individuals is reached who adopt different patterns of behaviour. Bilharz and Fricke (2011) illustrate this with the example of (auto-) mobility. The use of private cars reproduces and consolidates car-centric structures, which may be weakened if fewer people drive cars. Possible reasons for reduced car use, which demonstrate the interdependence between individual behaviour and structures, include: new laws might have increased the tax burden on petrol, new codes of signification, such as "car sharing is a good way to get from A to B", might have become common, new norms, such as "car driving is harmful for the environment", may have been established, or the structures of bargaining power may have changed because of a decreased number of jobs in the car industry[182].

Thus, this perspective strengthens the view that both the individual and the political levels are important in the realisation of sufficiency, as both are interdependent.

4.1.4 Individual and collective responsibility for sustainability

The preceding sections have discussed the benefits and limitations of individual and political approaches to sufficiency. In this context, references are often made to the notion of *responsibility*. For instance, Stengel (2011: 225 ff.) identifies the "delegation of responsibility" among different potential actors as one of several barriers that impede the mainstreaming of sufficiency (see also Winterfeld 1993, Fudge/Peters 2011: 802). Who is to blame for environmental problems, individual consumers, companies, or the policy level, and who is responsible for according solutions?

As already expressed in the quotation by Jürgen Dahl at the beginning of this chapter, there are some who think that it is unfair to make people believe "that they are personally responsible for present-day ecological dangers because they consume too much or proliferate too readily" (Bookchin 1989), and who diagnose a problematic "individualisation of responsibility" for environmental problems (Maniates 2002), which might be used as an excuse not to take effective action at the collective level (cf. section 4.1.1). As opposed to this, others point to the importance of "consumer responsibility"[183], insisting that consumers can make a difference

182 However, Bilharz and Fricke also emphasise that individual behaviour changes do not always have the potential to destabilise existing structures. Often, the characteristic of common goods problems leads to a situation where those who do not act in an "environmentally sound" way are rather rewarded with systemic incentives for not following the example given by others (Bilharz/Fricke 2011). To remain with the example of cars, if some people voluntarily give up their private car, more parking space becomes available to those who own a car, which makes car use again more attractive.

183 This notion was probably used for the first time by Hartley Withers as early as in 1920 in a remarkable article called *The Consumer's Responsibility* (Withers 1920), although in a different context. Withers was concerned about the poverty of workers and urged middle-class consumers to cut their extravagant and luxury consumption to enable investments which would allegedly help to produce necessities for the workers. "Consumer responsibility" is also often contrasted in the

with regard to sustainability (e.g. Bilharz et al. 2011, Heidbrink/Schmidt 2011, Schrader 2011)[184].

There are thus diverging opinions about whether responsibility can be assigned to individuals for the adverse environmental impacts related to their consumption decisions or whether this would overburden them due to the complexity of the issues involved and the nature of cumulative effects. At first sight, this does not seem to have too much to do with my argumentation here: I argue that because individual action is not effective enough, something (in the sense of sufficiency) needs to be done at the collective or policy level as well. However, as these things—what is effective and who has the responsibility to act—are often conflated in the discussion, I think it is worth trying to clarify the term responsibility as well as the arguments related to consumer responsibility.

"Responsibility" is an intensely discussed notion with regard to ethical and legal questions. Two main meanings of the term are usually distinguished between. In the first one, it means the "competence" or "area of responsibility" of someone who performs a particular function for something. The second meaning is responsibility as the attribution of actions and their consequences to someone or something (imputation).[185] This second category can be further split up. When someone refers to responsibility in ethics, this usually means moral responsibility. Moral responsibility is different from the mere causality of an action or event and its consequences, for example the weather in spring and the according development of the vegetation (Eshleman 2009). To distinguish moral responsibility coarsely from these other meanings, Eshleman (2009) suggests that the following description is sufficient: "to be morally responsible for something, say an action, is to be worthy of a particular kind of reaction—praise, blame, or something akin to these—for having performed it."

The matter dealt with here is not a particular blameworthy action which directly and on its own leads to a certain outcome. Rather it is an accumu-

literature on consumption with "consumer sovereignty", the idea of the autonomous and perfectly informed consumer exercising consumption decisions according to his or her preferences (e.g. Princen et al. 2002).

184 Cf. also the other contributions of a series of articles in the journal GAIA related to this subject, triggered by an article by Armin Grunwald dealing with "the privatisation of sustainability" (Grunwald 2010, 2011, Bilharz et al. 2011, Heidbrink/Reidel 2011, Petersen/Schiller 2011, Siebenhüner 2011).

185 In German: *"Zuständigkeit"* (competence) as opposed to *"Zurechnung"* (imputation) (cf. Wieland 1999: 7).

lative problem: even if one individual releases an enormous amount of greenhouse gases or ozone-depleting substances due to his or her consumption habits (say, 20 times the "normal" per-capita level), this alone would be harmless on a global scale. However, if many individuals do the same, the effects accumulate and then are likely to produce adverse impacts for everyone (cf. section 4.2.2).

Thus, there are various potential actors that could be held responsible: individuals, collectives, and other actors, such as those who contribute especially large shares to the problem (e.g. large companies). As regards individual responsibility, some authors have stated simply that these things are too complicated for individuals and that they are not in a position to take action effectively, so moral responsibility cannot lie with them (e.g. Spaemann 2007: 50, Petersen/Schiller 2011: 159). Such statements seem to conflate two different questions that should rather be dealt with separately: on the one hand, whether individuals do have moral responsibility for (their share of) adverse environmental impacts of consumption and, on the other, at which level action can be taken most effectively in practical terms. The term "responsibility" is often used in a wrong way in this context. From the assumption that individuals are not in a position to influence the outcome significantly by changing their habits, it is derived that they do not have the responsibility for (their share of) these outcomes, or that their responsibility covers only their acts as "citizens" which aim at influencing the policy level (e.g. Spaemann 2007: 50, Petersen/Schiller 2011: 161). This conclusion is overly simplistic. There is substantial debate in environmental ethics about whether individuals still have a responsibility to reduce their contribution to accumulative problems (see e.g. Johnson 2003, Caney 2005, Sinnott-Armstrong 2010, Baatz 2014), and opinions about this are diverging. They include views such as that individual consumers have no responsibility and no obligation at all with regard to accumulative effects, but also that individuals do have the responsibility and according obligations to reduce, for example, their share of greenhouse gas emissions towards sustainable per-capita levels "as far as can reasonably be demanded of them" (Baatz 2014: 10).

As regards the option of assigning responsibility to groups, which is more relevant for my subject here, there is controversial debate about whether responsibility, including from various aspects of consumption[186], can be assigned to collectives and companies at all. Some terms used in this sense include "collective" (e.g. Kutz 2000, Cripps 2011, Smiley 2011, Schwenkenbecher 2013), "cooperative" (Nida-Rümelin 2007), and "shared" (Young 2004, 2011) responsibility, while earlier liberal theorists tended to deny the possibility of assigning responsibility to groups (e.g. Hayek 1960: 83). The possibility of assigning moral responsibility to collectives is questioned particularly because it is unclear whether groups as such are at all capable of acting or acting intentionally, as they, unlike individuals, do not have proper minds of their own (cf. Smiley 2011). However, these concerns are usually related to group-based harm that is not merely accumulative, such as crimes committed during wars. With regard to accumulative problems, there is widespread agreement that collective responsibility makes sense as an aggregative phenomenon (Smiley 2011).

To that effect, there is also widespread agreement that responsibilities for accumulative environmental problems, such as climate change and the obligation to find according ways to remediate them, exist at the collective level (cf. also section 4.2.2). However, it is still a different thing to say that since individual, uncoordinated acts are not sufficiently *effective* in solving accumulative environmental problems (as pointed out in section 4.1.1), solutions should be looked for particularly at the collective level. In addition, as stated above, this does not mean that individuals (or other actors) might not still be held responsible, too, for their share in these problems, and that there may be according individual obligations.

4.1.5 The alternative: moralising individually harmless acts?

As an alternative to collective, binding measures in order to change consumption patterns, as this might seem too difficult, society might try to disseminate the attitude that certain consumption acts are morally condemnable and start "naming and shaming" those who do not act in line with this attitude.

Rather as a side remark, Sinnott-Armstrong states the following at the end of a text in which he argues that individuals have no obligation not to

186 For instance, the responsibility for the conditions in sweatshops producing cheap clothes (Young 2004, 2011) or emitting greenhouse gases (Caney 2005, Cripps 2011, Schwenkenbecher 2013).

drive a "gas-guzzling" SUV just for fun in order to avoid contributing to global warming (Sinnott-Armstrong 2010: 343)[187]:

> *"Even if individuals have no such moral obligations, it is still morally better or morally ideal for individuals not to waste gas. We can and should praise those who save fuel. We can express our personal dislike for wasting gas and for people who do it. We might even be justified in publicly condemning wasteful driving and drivers who waste a lot, in circumstances where such public rebuke is appropriate. Perhaps people who drive wastefully should feel guilty for their acts and ashamed of themselves, at least if they perform such acts regularly; and we should bring up our children so that they will feel these emotions. All of these reactions are available even if we cannot truthfully say that such driving violates a moral 'obligation'. And these approaches might be more constructive in the long run than accusing someone of violating a moral obligation."*

However, it is highly questionable whether such an atmosphere of public rebuke and moralisation towards legal and individually harmless acts would be preferable in a liberal society. Other authors have expressed deep reservations about stimulating public pressure on those whose consumption patterns are not in line with sustainability (especially Petersen/Schiller 2011; see also Grunwald 2010, Hellmann 2011). As Petersen and Schiller (2011: 160 f.) argue, this is an instrument used primarily by certain illiberal regimes which politicise "the masses" in order to label intrinsically harmless private activities, for instance, as "sabotage to building Socialism" and prosecute them. This "coerced voluntariness", according to Petersen and Schiller, is much more dangerous than any formal coercion by means of the law, against which one can fight with juridical means, which is not possible in the case of public pressure. They conclude that freedom is not threatened by the rules that a "strong state" enforces, but much more by a weak one that relies on mobilising public pressure against individuals in their private consumption decisions for those aims that it cannot achieve by means of legislation and enforcement (Petersen/Schiller 2011: 161). Another problematic consequence of public pressure not mentioned by Petersen and Schiller is that people might react with defiance instead of the behaviour intended.

---

187 In this text, Sinnott-Armstrong actually endorses the idea that the state should come up with regulations to solve the problems related to climate change, but suggests this—as I said, at the very end of the text—as an additional way to deal with individual behaviour.

I tend to agree with Petersen and Schiller; however, I think the issue is more of a tightrope walk than they admit. It seems reasonable for the state to run, for instance, information campaigns in order to show people the consequences of certain consumption patterns (cf. section 5.1), which might already be perceived as a sort of pressure or provide a reason for some individuals to put pressure on others.

The ways in which the state might try to make individuals change their consumption patterns, as individual-level sufficiency is not effective enough, will be further explored in the remainder of this chapter.

4.2 Justification of state interventions to protect the environment

4.2.1 Justifications of political interventions in a liberal society—the harm principle

Despite the emphasis on the individual's liberty within liberalism, there is consensus among liberals that there must be ways to restrict it, even though some say that real liberals should keep them as narrow as possible (Feinberg 1984: 14). In particular, a well-known rule allows for such restrictions, which all variants of liberalism can agree on—the "harm principle", famously phrased by John Stuart Mill in his essay *On Liberty* (Mill 2009 [1859]: 32-34):

> *"[...] the sole end for which mankind are warranted, individually or collectively, in interfering with the liberty of action of any of their number, is self-protection. That the only purpose for which power can be rightfully exercised over any member of a civilised community, against his will, is to prevent harm to others."*

This principle can be split up into a positive and a negative element: the positive stipulates that it is legitimate to interfere with individual liberty if the purpose is to prevent harm to others; the negative element states that this is the *only* reason for which power can be rightfully exercised[188]. In the context of this study, I am particularly interested in the positive claim.

Besides the harm principle, other principles related to limiting liberty have been proposed. Feinberg (1984: 26 f.) presents a list of these, including for instance a "benefit-to-others principle", paternalism, and perfec-

188 Libertarians in particular endorse this negative element of the principle.

tionism (see Box 6). In practice, many types of interference in modern states can be justified only on the basis of any of these other principles.[189] Even Mill did not reject all of them as consistently as he pretended to do[190]. Some of these principles would also be suitable to make the case for state intervention with individual choices for the purpose of environmental protection, and attempts in this sense have been made by other authors (Box 6).

However, none of these other ideas enjoy such widespread acceptance among liberals as the harm principle. Therefore, if interference in individual consumption decisions can be defended on the basis of the harm principle, it should be broadly acceptable.

Box 6: Liberty-limiting principles[191] that have been proposed in addition to the harm principle and their potential application to the environmental field

Benefit-to-others principle/Utilitarianism: according to a benefit-to-others principle, not only avoiding harm to others, but also realising potential benefits for others are a good reason to interfere with individual liberty. This can be interpreted as including the utilitarian view, which aims at maximising overall benefit. Even though it has been criticised substantially because of its potential to justify the denial of even the most fundamental rights to some people if this entails a higher overall utility, some modern theorists try to reanimate utilitarianism as a basis for public decision-making (e.g. Goodin 1995), and it is also part of the foundation of modern liberal theories of justice (Kelly 2005: 33). Actually, many political decisions are based on overall welfare maximisation, considering the common practice of carrying out cost-benefit calculations before adopting a policy. Arguments from the area of "happiness economics" (cf. e.g. Frey/Stutzer 2002, Easterlin 2001) are also based on

189 In fact, the harm principle cannot even justify the important liberal norm of equal treatment (cf. section 2.1.2), as Linklater points out (2011: 90): "There is nothing in the harm principle that requires that people should be treated as equals within the same community. A quite independent conception of equal human subjects has to be grafted onto the harm principle before it can contribute to progressive politics of that kind. From that vantage-point, the commitment to equality is at least as, if not more important than, the harm principle."

190 See e.g. the analysis in Brink (2008: section 3.6). With regard to utilitarianism, it is obvious that Mill was a fiery supporter of that idea. Further, Mill is sometimes considered to have been a perfectionist, too. For a discussion of the potential contradictions resulting from this, see e.g. Schefczyk (2011).

utilitarian ideas and are sometimes used by environmentalists to criticise the growth economy. In relation to consumption patterns, it could be argued that utilitarianism is a well-suited basis for limiting consumption in rich countries, as foregoing a unit of "trivial" consumption in a wealthy society for the sake of a higher level of basic need satisfaction in a poor society would increase overall utility. However, this principle still appears to be much weaker and be less accepted than arguing via harm, which is why I do not discuss utilitarianism in detail here.

Perfectionism: on a perfectionist account, policies should not be neutral (cf. section 2.1.3) but designed to advance some ideal of the good life. Ideas of this kind are popular among environmentalists and there are a number of proposals aimed at limiting environmentally unsound consumption and, at the same time, making it easier for humans to realise a certain vision of the good life (e.g. Skidelsky/Skidelsky 2012, Schneidewind/Zahrnt 2014). There is much controversy about the compatibility of perfectionism as such and liberalism, as outlined in section 2.1.3. For a discussion on this issue in relation to environmental issues, see sections 4.4.1 and 4.4.3.

Paternalism: paternalism is defined as interference in the liberty of another person against their will, where the act of interference is motivated by the assumption that the person will be better off or protected from harm (Dworkin 2010). Acts of paternalism are explicitly excluded from Mill's harm principle by it referring only to harm *to others*. Regulations that require, for instance, people to wear a seatbelt when driving a car are often seen as paternalistic, although it could also be argued that they aim at reducing the social costs of accidents. As paternalism targets specifically individual well-being, its application in the environmental realm is rather uncommon. Nevertheless, additional benefits for individual well-being are often promised to individuals when they change their behaviour in the way environmentalists have in mind, for instance as regards their health (e.g. when riding a bike or adopting certain nutrition habits).

Feinberg (1984) further lists the principles of *offence*, which aims to prevent serious offence to others (as opposed to harm), and *legal moralism*, which prohibits conduct that is "inherently immoral" even though it causes neither harm nor offence. These, as well as other derived principles listed by Feinberg, are of rather limited interest for environmental questions and are thus not dealt with here.

191 Mainly based on Feinberg (1984: 26 f.). Note that Feinberg's area of application is the criminal law. His analysis can therefore only be applied to policy in general to a limited extent and requires some adaptations.

The harm principle has been discussed extensively in philosophical literature. In particular, the notion of "harm" is controversial[192], as its interpretation determines when state interference with individual freedom is legitimate. Kernohan puts the problem in a nutshell as follows (1993: 51):

> *"The problem with applying Mill's principle is determining what is to count as harm. If we count mere hurt, offence, annoyance, and mental distress as harms, the principle will countenance political interference with nearly every activity, and liberty will amount to naught. On the other hand, if we count only physical damage to persons as harm, most every activity will be permitted and there will be little scope for the political protection of persons."*

In this regard, a reference to Feinberg's account of harm is often made. Feinberg (1984: 31 ff.) defines harm in the sense of the harm principle as setbacks to interests that are wrongs at the same time, that is, they violate other people's rights. The reference to wrongs is necessary to exclude setbacks to interests that do not violate rights, such as losing in a fair competition. To find out what "setbacks to interests" are, Feinberg suggests applying a test as to whether the interest under concern is in a worse condition after the alleged setback than it would have been otherwise (cf. also Raz 1986: 414).

Defining harm by reference to interests, wrongs, and rights, however, seems to shift the problem of concepts in need of definition only to other terms. A sound account of harm must therefore give an indication of setbacks to which interests should count as harm (Lin 2006: 927). Feinberg tries to refine his analysis by differentiating between types of interests (1984: 73 f.). His focus is on "welfare interests", the "basic requisites for well-being" which are necessary to achieve ulterior interests. This includes, for instance, health, stability, and a certain degree of freedom from coercion. Setbacks to these interests are, according to Feinberg, the most serious but not the only possible harms. As regards wrongs, if they consist in what is prohibited by law, this implies that evaluating whether something counts as harm depends on its current coverage in the law[193]. This

---

192 A further discussion on the harm principle concerns the question of which acts are to be considered as merely self-regarding and which ones as other-regarding (cf. e.g. Schefczyk 2011: 30). It is sometimes claimed that there are no purely self-regarding acts which significantly affect the agent alone and no other people (Gray 1996: 49). This discussion is, however, less relevant here as the harms in question—environmental harms—clearly have effects on others.

193 Feinberg's focus is the criminal law, which cannot be effective retroactively, so in his account it makes perfect sense to define harm in terms of wrongful acts.

would mean that no new harms could be identified and that harms not dealt with in the law, for instance because of their transnational character, would not count as such. Lin (2006: 925) therefore suggests withdrawing the element of wrongfulness from the definition of harm, in particular for the realm of environmental law.

Indeed, it has been argued that it might be impossible to provide an objective or neutral definition of harm (Raz 1986: 414, Gray 1996: 135, 140, Lin 2006: 921, Linklater 2011: 89). Gray (1996: 140) has stated that

> *"determining when restraint of liberty is justified turns out to be an intractably difficult business, since it hinges on assessments of the relative severity of harms that, in their dependency on disputed conceptions of the good life, are inherently controversial."*[194]

Nevertheless, the lack of objectivity of probably any definition of harm does not render the concept useless (Gray 1996: 128 f., Lin 2006: 930). Lin suggests that harm should be conceived of as a normative concept and relate only to setbacks to *significant* interests, which depend on community norms about the interests that matter (Lin 2006: 984). What is "significant" therefore seems to be influenced by what a majority of the population considers as such[195]. In addition to this rather qualitative characterisation, the severity of a setback must be taken into account, as many types of setbacks appear to be problematic only when reaching a certain level. An example of this would be the intensity with which someone is beaten: this can range from a soft pat to a knockout stroke. Therefore, appropriate thresholds for harm have to be defined. The qualitative element in the definition of a specific harm thus needs to be supplemented by a quantitative one. In any case, some harms are rather undisputed, for instance those that are well beyond some perceived threshold, while others are more contested, either as such or because they are closer to the threshold.

As most authors writing on the issue of harm note, an instance of harm does not create an automatism that regulation must be set in place (Gray 1996: 128, Lin 2006: 901 f., Brink 2008: section 3.6, Cripps 2011: 185, Coglianese 1998: 46). Mill already stated that if someone affects the interests of others, this just opens the discussion about possible interference

194 Cf. also Miller (1976: 131 ff.), who sees harm as a subjective notion depending on individuals' life plans.

195 Cf. the definition of harm used by the German Advisory Council on Global Change (WBGU) as an impact of human activity or any other event, which is intuitively viewed as negative by a majority of the population (WBGU 1998: 36 f.).

and there may be good reasons not to interfere (Mill 2009 [1859]: 38, 214). Once a new (alleged) case of (potential) harm is detected, it depends on a number of case-specific factors whether the harm should be stopped and where to set the threshold, and whether state interference is justified and what form it should take (Lin 2006: 901 f.). This must also take into account what interests are at stake on the other side, and thus include an assessment of the costs of stopping the harm. The more serious the adverse impacts on the other side would be, the more substantial the reasons for interference must be (Coglianese 1998: 47).

The subject I am dealing with here, harm through environmental degradation, is one of the rather clear cases where an instance of harm is often quite obvious. To give an example, heavy air pollution clearly damages the interests of those concerned, as it impinges upon their health and violates their right to physical integrity. Still, there remains the problem of setting thresholds for the quantitative element in the definition of harm, that is, where to draw the line between harmful and harmless air pollution or other forms of degradation (cf. section 2.3.4). Of course, there are also cases where it is not clear whether an activity or a certain level of pollution will have harmful consequences or not, for instance in the case of a new substance that is released into the environment (say, a nanomaterial or pesticide), because scientific evidence is not conclusive yet. This, however, does not relate to problems in the definition of harm, but rather points to the question of how risk and uncertainty should be dealt with (cf. Box 2). In general, risks are also covered by the harm principle according to Mill: he mentioned not only damage to others, but also a "definite risk of damage" as a legitimate trigger for society to deal with the issue under concern (Mill 2009 [1859]: 232). The issue of risk and uncertainty in relation to environmental harm will be raised again below.

Finally, it should be mentioned that the forms of harm in question here are particularly those that can be traced back to human activities. For instance, natural disasters that are clearly not caused by humans, such as earthquakes, also constitute sources of harm; however, it would not be possible to find someone morally responsible for them (cf. section 4.1.4). Of course, potential harm due to a looming natural disaster may be a good reason for government interference in order to limit the harm that may occur, but this is not the focus of analysis here.

4.2.2 The harm principle and environmental impacts—accumulative harm

Given that the harm principle is a rather simple idea, apart from the difficulties with its interpretation, cases where it applies to the environmental realm can be constructed easily. Actually, it is the most appropriate justification for many environmental regulations (Lin 2006: 922).

To give an example, let's imagine someone who wants to construct a waste incineration plant right in the middle of a residential area, applying outdated technology and not using any filtering installations. The ground on which the plant is to be constructed belongs to this person. If there were no land use plans and no environmental regulations, nothing could prevent the construction of the waste incineration facility. Now the harm principle comes in. It is obvious that the action of the potential plant operator would directly harm the residents of the area by polluting the air with hazardous emissions, maybe including substances that cause severe coughing and are suspected of producing cancer. Without the plant, the residents would not suffer these harms. It is therefore clear, according to the harm principle (and in line with the moral intuition of most people), that society can (and should) legitimately intervene in this case and prevent the harm by prohibiting the construction of such plants in certain areas and imposing technical standards on the operators of waste incineration plants to limit environmental pollution.

Three things are important to notice here: first, that the causal chain that leads to the harm is quite transparent, given the scientific evidence related to such cases we have today; second, that the person or entity responsible for the harm can be determined easily; and, third, that the action of one individual or a single entity quite directly leads to the harm.

However, many cases of environmental impact are not so simple. Since the 1980 s, political philosophers have been discussing how to deal with activities which only produce harmful pollution if a large number of individuals contribute to them independently and in an uncoordinated way. As mentioned before (section 4.1.4), the resulting harm has been characterised as accumulative harm. According to Feinberg (1984: 229), such harm has five characteristics in common, which deserve a full citation as they raise a number of problems that will be of importance later in this chapter:

> *"1. A threshold of harm is approached, reached, or exceeded through the joint and successive contributions of numerous parties.*

2. These contributions are uneven in amount, and unequal in degree of care and in social value.
3. In respect to the harm of pollution, each contribution is "harmless" in itself except that it moves the condition of the environment to a point closer to the threshold of harm.
4. When these accumulations cross the harm-threshold, they constitute public *harms in that they set back vital net interests shared by almost everyone.*
5. Most of the activities that produce these contributions toward pollution are so beneficial in other ways that if they were to be prevented entirely, as a group, the resultant harm to the public would be as great or even greater than the harm they now produce."

Typical examples of accumulative environmental harms are car driving (say, without a catalyser), climate change as a whole, or the release of ozone depleting substances into the air. The harmful effects may fall back on the very same individuals who have caused them, but they may also be shifted in space and/or time and affect human beings in neighbouring regions or other parts of the world, or future generations.

Problems related to accumulative harm

That accumulative harms are harms in the sense of the harm principle is undisputed, as the consequence (the harm) is the same, regardless of whether it results from a single or various uncoordinated acts. However, it is much more difficult to find appropriate ways to deal with cases of accumulative harm, since the causal chain for the harm is not exactly traceable. To use Kernohan's example, in the case of stratospheric ozone destruction, which increases the probability of someone getting skin cancer, we cannot know whether the chlorofluorocarbon molecule released by someone has destroyed a particular ozone molecule which could then not prevent exactly that part of the UV radiation that produced someone else's skin cancer from reaching the Earth's surface: "The problem of imputing [accumulative] harms to individuals arises because of our present, and most likely permanent, lack of knowledge of the micro-story." (Kernohan 2000: 345).

Because of this, it is more difficult to apply the claim contained in the harm principle that society may legitimately interfere with an individual's freedom of action in order to prevent harm. Kernohan (2000) has analysed in detail different options to impute the consequences of accumulative harms to individuals, including imputing the average consequence per share of individual contributions, abstract marginal consequences, that is, the difference a contribution makes to the total harm (at the moment when this is analysed), indexed marginal consequences, that is, the marginal consequence at the point when the contribution was actually made, and expected consequences, that is, using the probability that a consequence is produced by a certain contribution.

Kernohan rejects all of these imputation methods as unfair or implausible (cf. also Miller 2008: 130 ff.). His concerns are, among others, the non-linearity of many accumulative harms, threshold characteristics (including both lower thresholds up to which no harm is produced and upper thresholds beyond which all the harm that can be done is done), epistemic problems about individual contributions and their related marginal consequences, and the differences in the avoidability and desirability of the activities that produce the harm. He thus concludes that, except in some special cases where the analysed methods of imputation may be the best way of dealing with a problem, pinning down accumulative harms to individuals usually does not work. Still, he insists that this is no reason not to deal with accumulative harms. By giving an example of two cases with exactly the same harmful outcome, which was, in the first case, caused by an individual and, in the second, by various individuals, he reasons (Kernohan 2000: 364):

> *"Why should it be relevant to our need to prevent the harm that the first harm was the consequence of many similar acts by one individual, while the second harm was the consequence of many similar acts by different individuals? In dealing with accumulative consequences, we must, I think, give up on trying to impute them to individuals. Instead, we must attempt to promote accumulative benefits or prevent accumulative harms through systems of regulation, incentives and education that are workable and fair, but which do not depend on knowing who did what to whom."*

In line with this, Cripps (2011), for the example of climate change, suggests circumventing the focus on individuals by using an expanded understanding of collective responsibility (cf. section 4.1.4) and proposing a collective variant of the harm principle. She states that in dealing with accumulative environmental harm, the most appropriate way is to refer to the harm for which "we, collectively, are responsible" and establish related collective duties. The claim of weak[196] collective responsibility (1.) and

196 As opposed to "strong" collective responsibility, which refers to collectives that are formalised in some sense, "weak" collective responsibility in Cripps' sense applies to "putative" groups (cf. the claim and principle quoted below).

the collective harm principle (2.) defended by Cripps are as follows (2011: 174, 182):

> *1. "A number of individuals who do not yet constitute a collectivity (either formally, with an acknowledged decision structure or informally, with some vaguely defined common interest or goal) can be held morally collectively responsible for a harm which has been caused by the predictable aggregation of individual actions."*
> *2. "Curtailment of individual freedom, by a legitimate collective authority, is prima facie legitimate when the individuals are part of a group or putative group collectively responsible for harm, in the strong or weak sense, and the duties imposed on the individuals are fairly allocated as part of a collective endeavour to end that harm or compensate its victims."*

The responsibility claim takes effect, according to Cripps, only if certain circumstances are met: the individuals were aware of or could have reasonably foreseen[197] the harm that their aggregate behaviour would cause; they were aware that a sufficient number of others were performing similar actions that were adding to the harm; the harm could have been avoided collectively by acting otherwise (which they could have done without unreasonable demands) (Cripps 2011: 174 f.).

Cripps does not discuss in detail potential individual duties arising from the collective responsibility claim, but she mentions that these would have to satisfy certain conditions. As indicated in her collective harm principle, these conditions would, among others, include the fairness of the distribution of duties imposed on individuals in terms of, for instance, the costs of compliance, as well as the representativeness and legitimacy of the institution in charge of the task of imposing such duties (Cripps 2011: 188).

Typical excuses to the principle and lines of critique

A number of possible objections and excuses to a principle of accumulative harm have been raised with regard to both environmental and other forms of harm (cf. e.g. Kernohan 1998: 83 ff., Cripps 2011: 184 ff.,

197 On the question of knowledge about the harm that an action is likely to produce, see Miller (2008: 129 f.). Accordingly, responsibility can at least be assigned from a point in time at which there is sufficient scientific knowledge about probable environmental harms. Whether someone can be held responsible for actions back in time from that point is controversial, but anyway not my focus of analysis here.

Schramme 2011, Voget-Kleschin 2012: 107 ff.). In synthesis and ranked by increasing intricacy, these are:

(1) Consent of those harmed
(2) Lack of intention of those contributing to accumulative harms to produce these harms
(3) Unacceptable costs to avoid the harm
(4) Many harms will only happen with a certain probability and in the long run
(5) Individuals are held responsible because of the parallel actions of others and indirect effects
(6) It is too difficult to determine the individuals belonging to the group that causes the harm and to regulate their relevant activities adequately

These objections will be discussed one-by-one in the following and respective counterarguments will be raised.

As regards the first point, according to the harm principle, society should not interfere in individual action when those who are harmed have consented to the action (otherwise this would be regarded as paternalism). For cases of environmental harm, this is rather unlikely, except where those that benefit from an action are the very same people that are harmed (cf. Cripps 2011: 186 f.). However, most examples of accumulative environmental harm do not work in this way, since these harms often transgress borders and will also affect future generations, so consent cannot be taken as a general excuse.

As regards the second point, a lack of intent is commonly not accepted as an excuse to assigning moral responsibility in cases where harm was produced individually (e.g. Dworkin 2011: 290 f.). It is sufficient that someone could have expected that an activity would lead to harm, so intention should also not be a necessary condition for taking preventive action in cases of accumulative harm (Cripps 2011: 178, Kernohan 1998: 76 f.). Here it is worth mentioning again that wrongfulness is not a necessary element of activities that result in accumulative harm. It is thus necessary to distinguish between harm that gives reason to ex-post penalties, where intent or at least negligence are most often involved, and harm that gives society a reason to act and adopt mechanisms that will avoid (further) harm in the future. Once society has established such a mechanism, for instance in the form of law, further negligence may then be a reason for penalties.

Third, as regards the costs needed to avoid the harm, it has to be taken into account that the individual actions under concern are not harmful in themselves but may even be beneficial (Feinberg 1984: 229, Cranor 1985: 244, Schramme 2011: 343). The social costs might therefore be "unacceptably" high. However, the harms can also be unacceptable. In cases where environmental harms are caused mainly due to "luxury" purposes in rich countries and negatively affect mainly poor parts of the world, the vital interests of those harmed outweigh the costs in order to avoid the harm (Cripps 2011: 185; cf. Shue 1993). Further, the (accumulative) harm principle itself does not determine the character of the duties to prevent the harm. Prohibition of an activity that brings about accumulative harm is not necessarily the best solution, but there may be sufficiently effective ways of imposing a lesser burden on those benefiting from the activity. For instance, to keep the problem of accumulation of lead in the environment below the threshold of harm, leaded gasoline has been banned, not cars (Kernohan 1998: 86, cf. also Cranor 1985: 247 f.). Generally, the solution that is minimally onerous but sufficiently effective should be chosen, and it must be reasonable to those affected. This principle, and the issue of the costs of avoiding harm, will be dealt with again below (cf. section 4.3.5).

Fourth, the mere fact that a harm will only occur with a certain probability has, as early as in Mill's account of the harm principle (see section 4.2.1), not been accepted in itself as a reason for not applying the principle. Many environmental problems are caused by complex processes so that it is seldom possible to predict a certain harmful consequence with certainty. Rather, assessments of risk can be provided, but often uncertainties will remain (cf. section 2.3.4, Box 2). Thus, while not serving as a general reason for not applying the harm principle, cases of risk and uncertainty raise the question of what society can legitimately do to prevent potential harms, as these not only exist on the side of environmental effects, but also on the social side if a project is not realised or a technology is not used (cf. Box 2). Section 4.3.5 will again take up the issue of interference under conditions of risk and uncertainty.

In the same vein, pointing to the time that will pass until harms (potentially) unfold does not serve as an adequate excuse. Of course, the farther in the future a harm is supposed to occur, the greater the uncertainties involved. Also, it becomes less clear whether science will have developed better technical solutions to the problems under concern by then or not. However, if future generations are included in the assessment, which is a clear demand of any theoretical approach to sustainability (cf. section 2.3), potential long-term harms have to be taken into account (Voget-Kleschin 2012: 108). A possible answer to potential harm in the future is discounting it; the correctness and legitimacy of such calculations is, however, discussed extensively among philosophers and economists (cf. section 2.3.3).

Fifth, it might also be objected that the harm is not the direct result of individual action but of the indirect effects of uncoordinated activities, and that it would therefore seem that in taking collective action and cutting individual freedom, individuals are penalised for what others do (Kernohan 1998: 83) or, more precisely, because of the fact that others are performing the same intrinsically harmless acts.[198] However, it is not true that individuals are penalised for the actions of others, which would be against the standing of the individual within liberalism. They are only penalised for what they do (or are rather prevented from doing it again) within the context of a regulatory scheme that has been set up because others are doing the same and the accumulated actions would cause harm (cf. Kernohan 1998: 83, Feinberg 1984: 230). Of course, there is not only no intention to bring about harm; there is not even the intention to act collectively at all (Cripps 2011: 179). However, as soon as there is sufficient awareness of the problem and of the fact that others are doing the same, which might lead to crossing the threshold of harm, further denial of the need for a collective solution would be negligent (cf. Rees/Westra 2003: 116). As mentioned before, it is also unclear why the mechanism of the genesis of a harm should be generally relevant for the need to stop the harm (Kernohan 1998: 77, 2000: 364). The mechanism is decisive for the question of whether those contributing to a harm should be penalised, but not for the need to stop the harm.

Finally, in accumulative harm cases, it is often rather difficult to determine who is a member of the group responsible for the harm (Cranor 1985, Kernohan 1998: 84, Cripps 2011: 181). The "relevant community" will often not be a typical group of people from a clearly determined geographical area, but will rather consist of those individuals (and corporations or nations) contributing to an environmental problem through certain activities and may not remain constant over time (Cranor 1985: 251). Sometimes, membership in this sense is clearly defined, as in cases where individual contributions result from the same sort of activities and are similar in extent. However, in many cases contributions to harm can be sys-

198 Cf. e.g. Raz (1986: 416): "It is true that an action harms a particular person only if it affects him directly and significantly by itself. It does not count as harming him if its undesirable consequences are indirect and depend on the intervention of other actions."

temic and indirect, can result from various activities that are difficult to compare, or can range from tiny to substantial amounts. For this last possibility, Cripps (2011: 181) suggests that someone could become a member of the group when this person's individual contribution exceeds the amount generalisable for all without producing harm. Still, this does not answer the question of how to deal with effects resulting from very different activities. Picking only some activities that stand out, such as the practice of driving a "gas-guzzler" (cf. Sinnott-Armstrong 2010), and blaming them does not seem to be fair or effective. This issue will be discussed further in section 4.3.6.

It has become clear now that tricky cut-off problems may arise in cases of accumulative harm, and that it can be very difficult to achieve a fair distribution of burdens to adequately regulate the relevant activities. This is, however, no reason not to deal with accumulative harm, but requires a thorough analysis and solutions on a case-by-case basis (Cranor 1985: 251).

Conclusion on accumulative environmental harm

As a conclusion to the discussion above, it can be stated that collectives can be held responsible for accumulative environmental harms, and that political action is in principle legitimate or even required to stop these harms. Or, in the words of Ullmann-Margalit (1976): if too many people carry out an act *a*, which has undesirable consequences precisely because too many do *a*, this

> *"[...] imposes an obligation to* [sic] *the relevant community (call it C) of people involved in the situation to see to it that not every member of C does a. (Or, more precisely, that the number—or proportion—of C-members who do not do a is sufficient to avert the undesirable consequences that would ensue from everyone's doing a.) Roughly this means that the community is called upon to organize itself in such a way that the appropriate devices, such as division of labor or social norms, are generated so as to ensure that the harmful state of affairs where everyone does a does not come about."* (Ullmann-Margalit 1976: 516)

However, this does not mean that any kind of measures are always justified—a balance must be found considering the liberties that are at stake on both sides as well as the costs involved, and possible duties must be imposed in a fair way.

Independent of the progress of the philosophical debate on accumulative harm, there are a number of actual cases in the environmental realm where effective collective (including clearly coercive[199]) measures have been taken to mitigate such harm. A good example that has already been mentioned is that of destruction of the stratospheric ozone layer through ozone depleting substances. This problem has been tackled[200] through international agreements adopted in the 1980 s, which led to a staged ban on the substances under concern, which enjoys broad acceptance. There is, however, another example very similar in its mechanism in unfolding its effects globally and involving highly complex causal chains, which certainly does not belong to those cases where effective collective action has been taken so far, namely climate change resulting from anthropogenic greenhouse gas emissions. This is also one of the most iconic examples of accumulative environmental harm, which recent philosophical research in this field has focused on[201].

Therefore, the question arises: if the two examples of accumulative environmental harms, ozone layer destruction and climate change, are comparable, why has it actually been so (relatively) simple to tackle one and not the other? The answer is of a technical rather than a philosophical nature: in the case of stratospheric ozone depletion, less harmful substitutes to the substances could be developed for most applications, the costs of rearranging industrial processes and using the substitutes were bearable (Hammitt 2000, Vanner 2006), and the users were not affected significantly in their consumption behaviour (Deutscher Bundestag 2013: 451). As opposed to this, the case of greenhouse gases is much messier: in most cases, they cannot simply be substituted, as they are, among other things, an unavoidable product of every combustion process; the costs of abating the problem in a purely technical manner (by using alternative sources of energy and energy efficiency) are enormous; therefore, individual con-

199 On the notion of coercion, see section 4.3.5.

200 Cf. chapter 1.

201 Note that in the discourse on climate change, there are two approaches to stopping (or at least avoiding) harm: mitigation, that is, decreasing the emissions of greenhouse gases, and adaptation to environmental change. These two strategies could also be applied to other accumulative environmental problems, but my focus here is clearly on the mitigation approach as this corresponds more directly to the sufficiency strategy. Also, the question of to what degree adaptation is an equal strategy from the perspective of those harmed raises further ethical questions that lie beyond the scope of my discussion here.

sumers will very likely have to contribute to a solution to the problem by means of changing their consumption habits towards patterns that produce less greenhouse gases.

It is this last point that I would like to elaborate on in the next section. What is so special about consumption habits and levels that makes state interference in them so difficult and problematic? Why is it more problematic than interference in the form of technical measures that focus on the production of goods? Is it possible at all to see consumption levels as a cause of accumulative harms? These questions have, to my knowledge, not been discussed systematically in the relevant literature so far.

4.3 Consumption levels, accumulative harm, and interference

4.3.1 Consumption levels as the cause of accumulative harm?

If it is accepted that political measures are, in principle, justified in order to stop accumulative harm, it is still questionable whether this can be applied to certain consumption patterns that cause harmful environmental degradation, and if so, whether and how measures should be imposed to stop the harm. A number of authors have suggested that this should be the case; however, most of them merely scratch the surface of the argument. De Geus, for instance, states (2004: 95)[202]:

> *"It can [...] be argued that, even from a liberal point of view, it must be admitted that the highly valued right to 'freely act and consume' cannot be completely unlimited. Comfortable living, convenience and luxury are not legitimate in circumstances when they inevitably lead to the damaging of the natural surroundings of others."*

De Geus then goes on to briefly consider Mill's harm principle and concludes that "a good case can be made in support of the government's right to 'regulate' so as to achieve moderation and restraint among citizens, where their 'excessive demand and behaviour' is clearly at odds with the overall quality of life" (de Geus 2004: 99). It is, however, overly simple just to make such claims and refer to the harm principle.

---

202 Scc similar statements in Barry/Doran (2006: 252) or Müller (2009: 86 f.). Also cf. the more radical position of Rees and Westra (2003: 116): "[...] at the limits of biophysical carrying capacity, routine acts of non-essential consumption ultimately translate into acts of violent harm against the poor and racial minorities."

A thorough discussion of this needs to deal with a number of problems in addition to those that have already been discussed in the preceding general analysis of accumulative harms, dealing specifically with the aim of remediating accumulative harms through addressing *levels of consumption*, that is, by applying sufficiency as policy. This includes at least the following points:

- From a philosophical view, what is the difference between state interference by means of technical measures and sufficiency measures?
- When can consumption as such result in harm, that is, when is consumption "excessive"?
- Examples of accumulative environmental harm typically deal with harmful emissions. However, how should we deal with environmental problems not consisting in pollution but in the excessive exploitation of natural resources, which critics of current consumption patterns usually include in their arguments? Can such problems be seen as causing harm to other human beings?
- If the state is to "regulate" citizens' demand and use of consumption items, what would be the appropriate degree of interference, given that the activities are harmless or even beneficial in themselves, that the costs of harm prevention might be high, and that causality chains are long and there are considerable uncertainties involved?
- Further, if it is agreed that consumption should be limited, it might still be difficult to find suitable solutions. Thus, how and on what basis should limits to "excessive" aggregate consumption be translated into reductions of individual consumption?

4.3.2 What is the difference in terms of restrictions on liberty between technology and sufficiency policies?

In chapter 3, I differentiated between the approach of aiming at environmental protection via improving technology and the approach of influencing consumption patterns towards sufficiency, which was defined as "the reduction of demand or use of goods and services with high environmental impacts in order to achieve per-capita consumption levels that ensure emissions and resource use to stay within the environmental carrying capacity". While technology mainly affects production and is perceived by consumers, if at all, only because of (limited) price increases in products[203], sufficiency directly targets individuals and their consumption behaviour (cf. Table 4 in section 3.2.2). Typical environmental policies that focus on technology, which are to be implemented by manufacturers, are, for example, energy efficiency minimum standards that electrical appliances have to meet; the promotion of electric engine systems for cars; creating recycling systems that are able to transform post-consumer waste into useful material; or decreeing the substitution of a harmful substance in a product with others without relevantly influencing its properties for the user, as in the case of ozone-depleting substances.

While technology policies are quite popular in industrialised countries, it is often claimed that sufficiency policies are inherently anti-liberal and threaten "consumer sovereignty" (cf. Huber 1995 a: 129 f., Stengel 2011: 326; cf. section 4.1.2). Thus, a closer look at the difference between policies aiming at technical improvement and sufficiency policies with regard to their implications for liberty seems necessary. Potential differences could be grounded in two aspects of such policies: first, they could consist in a potentially higher standing of the freedom to consume as opposed to the freedom to produce and, second, they could be inherent in the fact that sufficiency aims to limit throughput, that is, it asks for quantitative restrictions, while the technical approach "only" requires the environmental properties of production processes and consumption items to be improved.

Production, consumption, and liberty

To get a better understanding of this issue, we may return to Mill's *On Liberty*. Mill himself does not use the terms "production" or (final) "consumption", but he does, in some instances, refer to the liberties of trade or producers as opposed to those of buyers. With regard to trade, he states that the action of selling goods always involves the interests of others and society, and thus may, in principle, be regulated (Mill 2009 [1859]: 268). Most questions related to trade, according to Mill, do not touch upon personal freedoms, but are linked to a different doctrine, namely that of free trade.[204] For Mill, it is also, in principle, "an evil" to restrict trade, but not

203 Whereas substantial and systematic price increases leading to a decrease in demand, including those due to taxation, should be seen as sufficiency policies.

204 Cf. the distinction between political and economic liberalism (section 2.1.4).

generally because of the liberties at stake but rather because the restriction might not deliver the desired results (2009 [1859]: 270).

As regards consumption, Mill is much more critical about the legitimacy of interference. He is clearly supportive of the idea of consumer sovereignty, objecting, for instance, to the restriction of the sale of alcoholic beverages or poisons, where the aim of the interference is just to prevent or make it at least more difficult to obtain a certain good. He states that these kinds of interference "are objectionable, not as infringements on the liberty of the producer or seller, but on that of the buyer", since "the State might just as well forbid him to drink wine, as purposely make it impossible for him to obtain it" (Mill 2009 [1859]: 270, 254). Another statement by Mill further underlines this priority of the users' interests over those of the sellers (2009 [1859]: 284):

> *"Almost every article which is bought and sold may be used in excess, and the sellers have a pecuniary interest in encouraging that excess; but no argument can be founded on this, in favor, for instance, of the Maine Law[205]; because the class of dealers in strong drinks, though interested in their abuse, are indispensably required for the sake of their legitimate use. The interest, however, of these dealers in promoting intemperance is a real evil, and justifies the State in imposing restrictions and requiring guarantees, which but for that justification would be infringements of legitimate liberty."*

However, this rejection of imposing restrictions on the liberty of consumers, as Mill's examples illustrate, only relates to consumption that might harm oneself. Mill did not consider the possibility that consumption might also cause harm *to others*. Reasoning about the legitimacy of paternalistic measures is beyond the scope of my analysis here. Still, it can be derived from the above that Mill generally rates the liberty of consumers as more important than the interests of the production side. Indeed, he seems to consider the satisfaction of individual preferences, for example by means of consumption, to be an end, while trade and production are mere means of serving general welfare.[206]

---

205 With regard to the Maine Law, Mill explains: "Under the name of preventing intemperance the people of one English colony, and of nearly half the United States, have been interdicted by law from making any use whatever of fermented drinks, except for medical purposes: for prohibition of their sale is in fact, as it is intended to be, prohibition of their use." (2009 [1859]: 250).

206 Cf. also the following famous quotation by Adam Smith from his *Wealth of Nations* (2005 [1776], Book IV): "Consumption is the sole end and purpose of all

However, Mill was (also) a utilitarian and this downrating of the interests of traders seems to be an expression of his consequentialist thinking. It is thus far from conclusive that liberals would generally support this view. To take an example from the context dealt with here, one could think of attempts by producers or traders to "promote intemperance" through advertising luxury goods or providing the option of paying for such goods in instalments. It is unlikely that current liberal theorists would opt for state interference in such corporate activities just because they are (allegedly) "only" in the interest of producers or traders.

Technical strategies, sufficiency, and liberty

The problem thus might not be related so much to the divide between the rights of producers and those of consumers. Indeed, the divide between technical and sufficiency measures cannot be clearly made by describing them as affecting only producers *or* consumers, respectively, even if the focus is on consumers (cf. section 3.2.2). The point rather seems to be that technical measures have a less significant impact on liberty in general.

For instance, the technical measure of prohibiting the sale of refrigerators that do not meet certain energy efficiency standards can be interpreted from the consumption angle as consumers not being allowed to buy such refrigerators any longer. However, it is likely that consumers hardly notice the difference to the situation before when no standard was in place, as only devices that perform better in energy efficiency remain on the market. For comparison, a potential sufficiency measure with regard to refrigerators would consist in prohibiting the purchase of refrigerators *entirely*, or beyond a certain size, or capable of cooling down below a certain temperature. This would certainly be a significant (and arbitrary) infringement on consumer freedom, which would invoke strong opposition.

However, this kind of policy is not what sufficiency advocates usually have in mind when they propose a policy of sufficiency[207]. As outlined before, most authors who make such proposals rather aim at enabling strategies and setting incentives (cf. section 4.1.2). Still, apart from enabling

---

production; and the interest of the producer ought to be attended to, only so far as it may be necessary for promoting that of the consumer."

207 Except maybe when it comes to certain products they consider to be superfluous, such as SUVs, patio heaters and the like.

strategies, sufficiency policies tend to entail deeper infringements of individual liberty, as they imply quantitative restriction and aim to trigger or require behaviour changes, while technical strategies "only" aim at improving or changing products and processes. Thus, policies aiming at sufficiency have to be designed more carefully bearing in mind their implications for liberty. This aspect will be taken into account in the further analysis (especially in sections 4.3.5, 4.3.6, and chapter 5)[208].

4.3.3 When is consumption "excessive"?

As the sufficiency debate is primarily concerned with individual consumption patterns, one might look at a certain individual's consumption behaviour and judge it, by means of rough comparison with our normal standards, to be far beyond what we consider appropriate (e.g. John Travolta running his private airport). However, as discussed in section 2.3.3 with regard to distributive justice, comparison with others and calling into question the actual distribution based on egalitarian reasoning is not the approach chosen here. We might also find that someone consumes too readily, apparently trying to compensate for personal problems by rewarding himself or herself with goods that are easy to obtain, but failing to achieve real happiness (cf. Brown/Cameron 2000: 28 f.). If one looks at environmental impacts, however, individual happiness is not the focus of analysis. In addition, since individual consumption acts usually do not as such harm others and the environmental problems under concern have been characterised as accumulative problems, final consumption by a particular individual cannot as such be judged to be excessive or not without being embedded into a broader perspective of macroeconomic resource consumption.

---

208 It should be noted that besides this, a difference between technical solutions and sufficiency that is likely to be a major reason for the disinterest in sufficiency on the political level concerns their potential effects on economic growth. While the promotion of technical solutions is perfectly in line with the paradigm of economic growth (cf. section 3.1.5), the impacts of widespread adoption of the sufficiency strategy on the economy are unknown (cf. section 4.1.2). However, as economic growth as such is neither a necessary prerequisite for sustainability nor liberalism (cf. sections 2.3.5, 2.1.4), potential adverse impacts on growth are not an obstacle to be dealt with from a theoretical perspective but rather when it comes to practical implementation and feasibility.

Thus, in a first step, the focus here is on the macroeconomic effects of consumption: total consumption is deemed "excessive" when the aggregate side effects of certain consumption activities produce relevant harms (or are likely to do so in the future) because they exceed the environmental carrying capacity. A judgement of when this is the case needs to be linked to a framework of criteria that helps assess when this excessive level is achieved.

In an article titled *Are We Consuming Too Much?*, Kenneth Arrow, Partha Dasgupta and a number of other renowned academics assess the macroeconomic level of consumption by applying two different criteria (Arrow et al. 2004). The first takes an economic vantage point and aims at maximising the present discounted value of current and future utility from consumption ("maximise present value criterion"), the second one is called the "sustainability criterion" by the authors and requires the maintenance of intertemporal social welfare over time. The analysis suggests that consumption seems to be too high if judged by these two criteria, although the authors state that the available evidence cannot be seen as conclusive (Arrow et al. 2004: 167; cf. also Ehrlich/Goulder 2007). However, the second criterion is a criterion of "weak sustainability", where losses of natural capital can be replaced by investment in other kinds of capital (cf. section 2.3.4), so it is no surprise that the article has provoked some reactions. In a reply to Arrow et al., a group of academics around Herman E. Daly ask: *Are We Consuming Too Much—for What?* (Daly et al. 2007). Instead of asking, "is consumption too large relative to investment?", they reframe the question so that it covers macroeconomic resource consumption in relation to the overall carrying capacity of the Earth: "Is the physical scale of total production[209] too large relative to the biosphere, even if it is efficiently allocated between the final uses of consumption and investment?"

My own understanding of "excessive" consumption is more in line with Daly et al. (2007) than with Arrow et al. (2004) and corresponds to the account outlined in section 2.3.4 on "critical" natural capital as well as my focus on the notion of harm[210]. Accordingly, I take macroeconomic consumption to be excessive if the aggregate side effects of consumption put

209 "Production", as Daly et al. use it, refers to what I call macroeconomic resource consumption (cf. Box 5, section 3.2.5).

210 Note that there are various other possible approaches to assessing whether consumption is excessive. Besides those suggested by Arrow et al. (2004), such an evaluation could be based on individual happiness (as mentioned in the introduc-

pressure on the environment beyond certain "safe" thresholds, which will (with a certain likelihood) lead to significant harm to other human beings[211]. Determining these thresholds is, of course, far from being an easy task, but nonetheless according attempts have been made, as described in section 2.3.4.

In a second step, the individual contribution that is still sustainable needs to be determined, that is, the average individual contribution to an existing accumulative problem from which no harm would be produced. This would, for instance, be the amount of emissions of a harmful substance an individual could emit which would be generalisable for all without exceeding the threshold of harm. Once the average individual contribution allowed is fixed, it constitutes a threshold beyond which continuing the intrinsically harmless activity would, in principle, become harmful. This is a common approach, above all, to greenhouse gas emissions. Care must be taken when applying this approach to any kind of environmental problem, as not all of them are of a global nature. For example, it would not make sense to fix a global threshold for individual fresh water consumption. Such thresholds, where needed, would have to be determined on the local level.

To say that there is an average individual contribution that is sustainable and which should not be surpassed is not the same as using this level in policies to set binding thresholds for individuals. It can also be argued that instead of equality, a different distribution principle should apply (cf. sections 2.3.3, 5.2.2). Or, as others have argued, a specific pollutant or resource should not as such be subject to any distribution principle at all, but rather be seen as part of a bundle of goods or resources, the fair allocation of which to individuals should be the aim of distributive justice (e.g. Hay-

---

tion to this section), or it could be linked to an egalitarian conception of distributive justice or, similarly, to the idea of strong sustainability, which demands that the stock of natural capital should not decrease over time (cf. section 2.3.4). Also, the level of consumption could be assessed by comparison to some level of needs fulfilment, and what is consumed above this level is deemed excessive. Or one could suggest a perfectionist account to assess consumption, such as that proposed by Skidelsky and Skidelsky (2012), or that by Sagoff (1997: 96): "We consume too much when market relationships displace the bonds of community, compassion, culture, and place. We consume too much when consumption becomes an end in itself and makes us lose affection and reverence for the natural world."

211 Cf. also Princen's notion of "overconsumption" (2002: 33).

ward 2007). From an environmental point of view, what is important is that the average contribution and thus the total are decreased to a sustainable level, but it does not matter whether some individuals pollute more than others. How to implement measures where macroeconomic consumption is "excessive" in order to curb individual consumption will be discussed further below (see section 4.3.6).

Finally, before saying that *consumption levels* themselves are the problem, it should be verified that there are currently no technical solutions available which will solve the accumulative problems of certain consumption patterns, or that technical solutions cannot effectively achieve a solution alone within a resonable period of time and with acceptable side effects and costs (cf. chapter 3). If this is the case, we can say that the consumption levels themselves are the cause of accumulative harm, and that consumption as such is excessive.

4.3.4 Can the excessive use of resources cause harm to others?

Two kinds of anthropogenic influences cause environmental problems: pollution of the environmental media air, water, and soil ("sinks"), on the one hand, and overexploitation of renewable natural resources[212], such as forests and fish stocks ("sources"), on the other (cf. e.g. Meadows et al. 2004: 51 ff.). Environmentalists see the high consumption levels of people in industrialised countries not only as problematic because of the related emissions and waste that go into the sinks, but also because of the exploitation of natural resources that accompanies these consumption patterns.

However, while it is clear how pollution of the sinks can result in harm to other human beings, this is less obvious in the case of overusing the sources in a globalised economy. For instance, with regard to the production of food, eating a lot of fish may contribute to the overexploitation of fish stocks somewhere, and the consumption of beef requires a much higher resource input (e.g. land use, fresh water) than the intake of the same amount of calories of other animal products, let alone calories from plants

212 Note that the extraction of non-renewable resources often produces environmental problems of the first kind—pollution—and is therefore not mentioned here as an additional category. Cf. chapter 1 (section on the "use of natural resources") and section 2.3.4, footnote 90.

(Eshel et al. 2014)[213]. However, is the high resource consumption of some the reason for making others fall below the threshold of needs and thus harming them because their provision of basic resources is not ensured any longer, or would the high consumption standards of some only be a problem if one aimed at a (more) egalitarian distribution of resources? And, if there is harm to human beings due to a lack of resources, does the harm really arise because of wealthy people's high consumption levels *as such*, or might the main reasons not rather lie elsewhere, such as in national economic and political conditions and certain immoral business practices, including violations of property rights (i.e. exploiting a certain resource somewhere illegally), bribing local officials to get access to a resource, and unfair contracts multinational corporations impose on local people?

In the 1970 s, *Limits To Growth* predicted that mankind would soon run out of non-renewable resources or that they would at least become extremely costly, and that the Earth would not be able to support mankind with sufficient food if the population and demand continued to grow exponentially (Meadows et al. 1972: 46 ff., 54 ff.). The authors' "world model" predicted that growth would, sooner or later, lead to absolute resource scarcity (cf. Box 7) and global collapse, which would obviously be harmful to the vast majority of people. However, the report was criticised substantially for underestimating the possibilities of productivity increase, searching for substitutes, and technical development in general[214]. Other theorists[215] have gone as far as claiming that mankind would never run out of resources because the empirical long-term trend has so far been that resources become ever more abundant (e.g. Simon 1998). Yet, since the late 1990 s, after a long period of decline, the prices of food, energy, raw materials, metal and mineral resources have been increasing steeply (UNEP 2014 a: 23 ff.; cf. chapter 1). There are now scientists who state that the business-as-usual scenario from *Limits To Growth* provides a surprisingly

---

213 To be sure, meat consumption and food production in general are also related to a number of sink-type problems, such as greenhouse gas emissions and water pollution (Eshel et al. 2014). However, I focus only on the exhaustion of resources as such here.

214 For a summary of the criticism, see Ekins (1993). It should, however, be noted that the criticism regarding *Limits To Growth* is itself also subject to dispute and has reversely been criticised as flawed and incorrect in many respects (cf. Turner 2014).

215 So-called cornucopians—cf. section 3.1.1, footnote 119.

adequate forecast of the actual development of most variables, and that global downturn may be imminent (cf. Turner 2014). While in the past the number of people in societies with high consumption levels was limited to a rather small global elite, now, with growing middle classes in large countries such as China, India, and Brasil, the pressure on global resources is much higher. In any case, extrapolating past developments of complex systems, such as the global demand and supply of resources, into the future is a daring venture, and the possibility of absolute resource scarcity cannot be excluded per se.

Today, the accusation that rich countries consume too many resources and thus undermine others' chances of meeting their needs is present, for example, in the criticism of their meat consumption, and, related to this, in the discussion on biofuel policies. Rich countries are accused of putting increased pressure on world market food prices because of their high demand for meat and financial support for biofuels, which require land that used to be used for food production. Therefore, prices are said to be increasing and less wealthy people cannot afford the same amount or quality of calories anymore (Oxfam 2008, AnimalAid n.d., cf. also WWF 2014 b).

It is obvious that someone will be harmed if they are deprived of the most important resources required for subsistence. If a person does not have access to a certain amount of food, this person will starve and suffer bodily and other harm. The required amount may vary slightly from one individual to another, but in principle there is an absolute threshold of daily calorie intake below which someone will be harmed in the long run if he or she cannot have this amount. Similar claims hold true for access to adequate clothing and shelter, for an environment which provides a minimum level of security, etc. (cf. Box 1 on "needs", section 2.3.3).

However, it is less clear why the high resource consumption of some should be the *reason* why others do (or will) not get enough resources to meet their basic needs. This would clearly be the case in a scenario where two castaways have landed on a lonely island, on which the only nutrients are delivered by coconuts, and one of them ate all the ripe coconuts (say, exceeding the amount of what that person needed) and let the other person starve. However, the real world is not so simple, as there are manifold other interplaying factors that have an influence on the (local) availability of resources, including local economic and political conditions, complex interactions in a global economy and world market prices including the possibility of financial speculations, other policies that have different, for instance, protectionist aims (e.g. subsidies), local pressure on resources due to increased population, temporarily unfavourable weather conditions, etc. Actually, a high demand for commodities is, in principle, even beneficial for poor countries that export these commodities (cf. Sagoff 1997: 92 f.), as high world market prices imply high revenues for them, at least temporarily. Thus, whether a high demand in wealthy countries has negative consequences for the poor or not also depends on the type of resource.

For the example of the causes of hunger, demand in rich countries is still one important factor driving world market prices of agricultural products and thus increasing the probability of local food shortages in poor regions. Currently, food is not scarce in an absolute sense at a global level, and hunger is rather an issue of its distribution (Voget-Kleschin 2012: 171). However, increased relative scarcity (cf. Box 7 below) on a global level can turn into regional absolute scarcity if basic resources are expensive enough that poor people cannot afford them any longer, and can lead to food riots (UNEP 2014 a: 25). It is also possible that with further increases in global demand for agricultural products due to further population growth, the growing demand for biofuels, and a dietary transition to an increased calorie and protein intake especially from livestock, food will indeed become scarce in an absolute sense (Voget-Kleschin 2012: 172 ff.).

Thus, even if the sheer "excessive" demand and consumption of wealthy consumers plays a role in absolute resource scarcity that harms the poor, there are numerous other political, economic, and environmental factors, as mentioned above, which can be more significant. Compared to accumulative harm through polluting the "sinks", it is somewhat harder to provide evidence of accumulative harm through excessive use of the "sources", since it is much easier to substitute a specific resource with other resources than to find substitutes for the essential functions provided by the sinks[216]. However, many sources are sinks at the same time: forests, for instance, are not only a source of wood and other resources, but also a sink for carbon dioxide. Therefore, their depletion can aggravate harms due to pollution as well. In addition, a great deal of resource use or extraction activities, for instance the extraction of non-renewable resources as well as many land use related activities, produce harmful pollution at the

216 Note that in the sustainability discussion it is sometimes suggested that, in principle, anything can be substituted (cf. section 2.3.4), including important functions such as the one provided by the ozone layer—with suntan cream. Similarly, one could imagine substituting clean air with breathing masks. However, I do not think these are acceptable (or even serious) proposals.

Box 7: Absolute and relative scarcity

Scarcity in general means that there is less of something than is being demanded. In economic terms, all things that have a price higher than zero are scarce, that is, not only is water scarce in desert areas, but also Rolex watches are scarce. However, from an environmental point of view (and not only from there), there is disagreement about such a broad definition of scarcity, as not all kinds of demands are seen as equally legitimate (cf. Daly 1991: 40, Scherhorn 1994, Winterfeld 2002: 28 f., Schramme 2006: 151 f.). Back in the past, the term was primarily used for times of hardship, especially with regard to food scarcity and hunger. Only with the establishment of classical economics did it become a notion linked to the infiniteness of human desires, and thus nearly all resources and goods were considered as scarce, even among very affluent people (Xenos 1987: 225, Hoeschele 2007). Thus, distinguishing between "relative" and "absolute" scarcity has been proposed (e.g. Daly 1991: 39). Relative scarcity corresponds to the definition used in economics mentioned above: things are relatively scarce if they come at a certain price, or, in other words, if something else must be given up to obtain them. In a relative sense, things are thus scarce in relation to other resources or goods and can be substituted (Baumgärtner et al. 2006: 489)[217]. As opposed to this, absolute scarcity is linked to the total availability of resources on the planet or in a region. This notion of scarcity is based on human needs rather than desires (cf. Box 1, section 2.3.3) and thus refers to the means required to satisfy these. Something becomes scarce in an absolute sense if there is not enough of it to satisfy basic needs, such as the need for food, shelter, etc., and cannot be substituted with other resources (Baumgärtner et al. 2006: 490). As Baumgärtner et al. (2006: 493) point out, in a particular context the two types of scarcity can turn into each other if one considers a different spatial scale (e.g. a country instead of a town) or if the empirical circumstances change, for instance if new technical possibilities of substitution are developed, new institutions emerge that foster exchange between different economic agents or social processes are organised in a more efficient way.

When I refer to the notion of scarcity in this study, I mean scarcity in an absolute sense, unless otherwise stated.

217 Cf. also John Rawls' notion of "moderate scarcity" (in which he follows David Hume), which seems to express the same as "relative scarcity" (cf. Hubin 1989: 185), and means that "natural and other resources are not so abundant that schemes of cooperation become superfluous, nor are conditions so harsh that fruitful ventures must break down" (Rawls 1972: 110).

same time (cf. chapter 1). Still, while polluting the sinks produces harm more directly, potential harmful effects linked to the overexploitation of some of the sources are more indirect. This has to be kept in mind when discussing potential policies of sufficiency.

4.3.5 What degree of interference is legitimate with regard to excessive consumption?

It was concluded above that collective action is needed to stop accumulative environmental harms, even when they are the result of "excessive" consumption, since individual voluntary action is not effective enough. Thus, political interference is, in principle, legitimate for this purpose. However, given that consumption acts are everyday practices that are harmless as such and become harmful only through accumulation and often complex chains of mechanisms, what degree of state interference would be appropriate to curb consumption is questionable.

Before dealing with this question, the term "interference" requires clarification. It is very broad and is usually not defined precisely. Mill uses interference or "interfering with the liberty of action" to describe the set of reactions that are, in his view, legitimate for society to deal with an instance of harm to others. It can be understood as an umbrella term for any purposeful influence the state enacts on individuals. Apparently in the same sense, Mill also uses notions such as "compulsion and control", including the means of "physical force in the form of legal penalties, or the moral coercion of public opinion", as well as "exercising power" and "coercion" in general (Mill 2009 [1859]: 32-34, 52; cf. Anderson 2014). Outside the scope of the harm principle, where no harm to others is expected, Mill considers the means of remonstrating and reasoning with someone, persuading, or entreating someone as appropriate to influence this person's behaviour (2009 [1859]: 34).

Of these terms, the notion of "coercion" is especially of interest, as it is a widely discussed term in political philosophy (cf. e.g. Hayek 1960: 20 f., 133 ff., Nozick 1969, Raz 1986: 148 ff.; see also Anderson 2014). In common language, the term sounds rather strong and negative, whereas "interference" appears to be more neutral. Coercion seems to refer to a specific kind of interference. If state interference in individual freedom can be presented as a continuum between interventions that have a strong influence on an individial's behaviour on the left end and those which only lightly influence the individual on the right end, coercion as a common sense notion is located on the far left.[218]

Within the contemporary philosophical debate, coercion is usually understood as an act that involves the *threat* of an unfavourable consequence in order to steer the coerced person's choice among the available alternatives of action (Raz 1986: 149, Anderson 2014). Only threats of significant negative consequences, which carry great weight for the coerced person, qualify as coercion (Nozick 1969, Raz 1986: 149). This suggests that with regard to the state, only certain types of interference should indeed be classified as coercion. However, theorists differ with regard to what they see as included in the concept of coercion (Kelly 2005: 52). Some hold that nearly all sorts of state interference are coercive, as even those that involve spending money rely on some prior coercive mechanism to get the money from taxpayers (e.g. Hayek 1960: 143 f.). As Robert Nozick has famously claimed, taxation of income from labour is therefore not much different from forced labour (1974: 169). As opposed to this, others do not even agree that taxation should be labelled as interference. For instance, Arneson argues that if the state were to offer opportunities to individuals, such as subsidising opera, this does not mean forcing but rather enabling them. Even if it is clear that someone must pay for this opportunity, this, according to Arneson, seems to "stretch the idea of interference beyond good sense" (Arneson 2003: 198).

It seems to make sense to distinguish directly coercive kinds of interference from others that involve spending money and where the coercion lies in a separate mechanism aimed to get that money. Only direct coercion will be labelled as coercion here. Interference by the state, then, is understood here as any purposeful influence the state enacts on individuals, with clearly coercive policy instruments at one end of the scale, and other instruments such as enabling strategies and information campaigns at the other end.

According to our moral intuition, the reasons for adopting clearly coercive measures must be more significant than those for less compulsory ones. This idea is embodied in the principle of commensurability, which is an important precept of public law, at least in German-speaking countries (*Verhältnismäßigkeitsprinzip*). This principle does not only require author-

218 Corresponding to the upper end of the categorisation of policy instruments as proposed by Jänicke (2003), cf. section 5.1.

ities to select only proportionate means to achieve a certain (legitimate) aim, but also to select the one which is effective and at the same time has the least significant adverse impacts on those concerned as well as on the general public from various possibilities.

Therefore, in search of appropriate solutions to accumulative harms caused by "excessive" consumption, one has to consider that rights and liberties are at stake not only on the side of those being harmed but also on the side of those contributing to the harm, as the harm in accumulative cases of harm is usually not intended (cf. section 4.2.2). Instead, it is a byproduct of an action that intends to achieve a different, legitimate aim. This would not be the case, for instance, if I hit somebody just to harm this person. In this case, there would be no reason to search for a solution other than penalising me and making sure that this does not happen again. As opposed to this, in accumulative harm cases the kind of interference chosen to prevent the harm from occurring has to depend not only on the severity of the accumulative harm, but also on the (not only financial) costs involved to avoid it. Still, in cases where the harm is severe and there is no other way to avoid it, prohibiting actions that lead to accumulative harm may be required despite the involved costs. Hence, coercion, too, is in principle a legitimate option with which to deal with instances of accumulative harm caused by consumption (cf. Cripps 2011: 183), if it passes the test of the principle of commensurability.

A further issue which should be taken into account in the degree of interference in consumption activities is that of risks and uncertainties (cf. Box 2). As stated before, environmental harms are often caused by long chains of mechanisms and consumption acts are often, particularly in the case of indirect impacts (cf. Box 5), far away from the place in the chain where these harms occur. Moving up along the causality chain from environmental effects to consumers amplifies the uncertainties involved, as there are more intermediate factors that influence the outcome (Wynne 1992: 117). Even if the presence of risk and uncertainty does not serve as a reason not to apply the harm principle (cf. section 4.2.2), it is still a factor to be taken into account in terms of the strength of measures taken to mitigate potential harms.[219]

219 Heinrichs (2013 a) argues that the perspective of risk focuses on specific negative consequences of certain paths of development, while in the sustainability context a broader perspective that provides positive alternative transition and adaptation processes appears to be needed. While I agree that sustainability as an idea is also

It is often argued that in cases of risk—defined as the product of the probability of an event of harm and its potential extent—cost-benefit analysis should be used to support decision-making (cf. Box 2). However, where the potential harm falls on other people than the benefits, such an approach might violate the rights of those who will potentially be harmed (Caney 2010: 170). As opposed to cost-benefit analysis, the principle of commensurability does take into account the differentiation between these two sides, while still addressing the costs (or benefits) on both sides, too. Interference in order to deal with cases of calculable risks to others can, based on the principle of commensurability, generally be dealt with like cases of harm to others. However, in cases of accumulative harm because of "excessive" consumption, a future outcome is often characterised by uncertainty rather than calculable risk. Uncertainty may be a reason to apply the precautionary principle, that is, to consider interference if an activity is suspected of causing significant harm, even if there is no full scientific certainty (cf. Box 2). Of course, in consideration of the principle of commensurability, the merest hint of uncertainty is not a reason to adopt disproportionate precautionary measures (WBGU 1998: 29).

As consumption refers to a wide and inhomogenuous set of activities, the appropriate degree of interference to deal with accumulative harm due to consumption depends on the concrete case under concern. As the previous discussion has pointed out, factors that should be taken into account in evaluating concrete cases include at least the severity of the (potential) harm, the costs on the side of those harming to avoid the harm, the assessment of different solutions available as regards the test posed by the principle of commensurability, and the risks and uncertainties involved.

4.3.6 How to translate limits on "excessive" aggregate consumption into ways to reduce individual consumption?

It was stated before with regard to "excessive" consumption that individual-level consumption can only be assessed within a scheme of macroeconomic consumption. If overall consumption exceeds certain thresholds, something has to be done about individual consumption in order to avoid

about providing such positive paths, my focus on avoiding harm and my minimum conception of sustainability (cf. section 2.3) requires me to adopt a rather narrow, indeed negative perspective.

harm because, in the end, it is individuals who consume and not abstract macroeconomic entities. However, how should this be done? The harm principle gives no answer to this, as it just provides a basis for taking action to stop the harm but does not entail any further principle of distributing the burdens among those who contribute to it.

Three potential solutions are identified in the following, which follow quite different approaches. The first targets the harmful pollutants and scarce resources under concern and is about limiting access to these and/or "getting the prices right". The second targets individuals and their contributions to an environmental problem and assigns "fair shares" to them. The third solution targets certain practices of consumption or consumption goods which are identified as problematic and establishes mechanisms to prevent people from performing these activities or purchasing these goods or at least to make them less attractive.

While the first two solutions are derived from economic and justice theory, respectively, the third is rather pragmatic and can be characterised as a "muddling through" approach based on a rough evaluation of which practices contribute a great deal to accumulative harms, while at the same time being of limited importance for meeting human needs (or could be replaced by other, less harmful practices). All three solutions have certain advantages, but are also flawed in some respect or will only work under certain circumstances. I will discuss them in general below, while more detailed examples will be provided later on in section 5.2.

Targeting harmful pollutants and scarce resources

To address problems at their root, it makes most sense to focus on the amount of resources that can be harvested and pollutants that can be emitted without crossing the threshold of harm. While this looks rather like a top-down supply-side solution, it also affects consumption: if the supply of a resource (say, redfish) is reduced (say, through fishing quotas) then the amount available for consumption will also decrease and consumer prices will, ceteris paribus, increase. In a similar way, one could tax a resource or emissions of a pollutant so much that demand for it will settle down at sustainable levels.

This solution corresponds to the typical answer of environmental economists, aiming to "get the prices right" for certain resources or pollutants, internalise external costs and leave everything else to the market.

Besides decreasing consumption of certain resources and emissions of certain pollutants to sustainable levels, this solution particularly aims to achieve an economically efficient allocation of resources and emission rights. Examples of policies based on this kind of instrument exist in many countries (cf. section 5.2.1).

While this is sometimes praised as a potential "catch-all" solution, it is not difficult to imagine how it can also generate significant problems, at least if the tax or restriction reaches a certain level. These include potential market distortions that would have to be carefully anticipated and remediated, such as the emergence of black markets or the overconsumption of possible replacement resources if a certain resource is taxed or a cap is introduced. In particular, price increases, if significant, could have harmful consequences on individuals who can no longer afford the good under concern if it is essential. This might make them fall below the threshold where basic needs are met. Examples of such cases are, for instance, discussed under the headline of "fuel poverty" as a result of the high prices of energy carriers (e.g. Boardman 2010).[220]

It seems therefore necessary to have a look at the distributive impacts of heavy taxation or caps on resource use or emission rights. Taxes on essential resources have to be moderate or accompanied by ancillary interventions in order to remedy potential harmful consequences. Setting a cap on the use of essential resources or emissions of essential pollutants seems to require a second regulatory "layer" in order to distribute the available amount.

Targeting individuals

From the vantage point of achieving a just distribution of the possibilities in oder to perform consumption activities that contribute to accumulative harms, it makes sense to focus on individuals and define a fair per-capita allowance to pollute[221] or use basic resources. A major advantage of this approach is that individuals are free to decide how to allocate their pollu-

---

220 For a more thorough discussion of environmental taxes, see section 5.2.1.

221 This term is to be understood very broadly, as the substance under concern in the most widely discussed case for such individual allowances, carbon dioxide emissions, is actually not a "pollutant" but a chemically rather inert substance which happens to influence the Earth's radiation budget.

tant and resources budget to their diverse consumption practices. The idea of establishing individual allowances to use a certain share of environmental carrying capacity was already expressed two decades ago in the idea of an "environmental space" (e.g. FoEE 1995): This concept is based on the simple idea that the maximal amount of a natural resource that can be consumed (or amount of a pollutant that can be emitted) without producing environmental harm should be distributed fairly among individuals, so that each is assigned a "fair share" (McLaren 2003, Hayward 2007). This share can be seen either as a restriction or, in a more positive sense, as the available environmental "utilisation space" that individuals can legitimately make use of (cf. Spaargaren/van Vliet 2000: 57).

Proponents of environmental space usually suggest an egalitarian approach, so that, according to them, the available capacity of the Earth to provide resources and absorb pollutants should be distributed equally among people within and between generations (cf. section 4.3.3). Alternatively, some suggest taking into account a historical "ecological debt" (e.g. McLaren 2003), which might result in a less-than-equal share for people who have in the past consumed more than their environmental space would allow. Critics of egalitarian justice would, however, ask why these allowances should be distributed equally and not according to a different mechanism, for instance need. Problems related to the distribution of environmental space will be discussed further in section 5.2.2 with regard to the example of personal carbon trading.

Further, there seem to be additional problems related to the idea of enforcing a policy based on environmental space. If we consider in detail the requirements and consequences of establishing such a system for a number of emissions and environmentally relevant resources, it quickly becomes clear what a bureaucratic nightmare this would be. Monitoring each individual's compliance with the limits of his or her allowed environmental space neither seems feasible nor desirable. This becomes even more obvious if one intends to include the resource use "embedded" in the production chains of consumption goods in order to cover indirect effects, too (cf. section 3.2.5 on consumption). However, not including these would endanger the idea of this solution as a particularly just one, as indirect effects can make up a significant share of the total of environmental impacts: in 2008, between 24% and 56% of environmental impacts (greenhouse gas emissions, land use, water use, material use, etc.) associated with consumption in the EU occurred outside EU territory, with increasing tendency (EEA 2014: 34 f.; cf. also chapter 1).

Indeed, proponents of an environmental space approach do not usually propose that a strict system to regulate individual consumption based on individual allowances is set up (e.g. FoEE 1995: 189). Instead, they see it as a useful approach for nations to assess the per-capita contribution of their citizens to accumulative harms, allow for a global comparison, and make according commitments. In addition, calculating a sustainable environmental space helps individuals to roughly assess their own contribution to global emissions and resource uses, and compare it with what is generalisable for all[222]. Therefore, individual allowances can mainly be a guiding tool, and supplementary approaches are needed in order to decrease resource use and harmful emissions.

Nonetheless, for the example of carbon dioxide emissions (or greenhouse gas emissions in general), establishing individual allowances as a policy is actually being discussed under the headline of "personal carbon trading" or similar names. This idea is sometimes praised as a sort of redeemer that is able to disburden mankind of all other "inefficient" solutions to carbon dioxide mitigation (e.g. Müller 2011). The example of personal carbon trading will therefore be discussed more in detail in section 5.2.2.

Targeting certain activities or consumption goods

In view of the problems linked to the two approaches, which are ideal representations of different theory strands, it makes sense to examine whether certain consumption activities or items that produce significant shares of accumulative harms could be targeted. Possible candidates include, for instance, using aeroplanes and private cars for transportation, buying hot meals "to go" in voluminous one-way packaging, or eating beef. Targeting such activities or consumption goods actually conforms to the prevalent political piecemeal approach, which is manifest in policies such as energy efficiency requirements for cars and appliances, or taxes on plastic bags.

This approach is sometimes perceived as arbitrary and unfair, as it seems to pick out certain outstanding practices that are accused of being especially environmentally harmful—such as the practice of "driving a

222 For instance, in internet tools that help to estimate how many planets would be needed if everyone were to have a lifestyle like oneself.

gas-guzzling sport-utility vehicle for fun on a beautiful Sunday afternoon" or the use of traditional lightbulbs (cf. Sinnott-Armstrong 2010, Schramme 2011: 345 ff.).

Arbitrariness should, of course, be avoided—the easiest and fairest way to do so would be to let individuals decide for themselves which consumption activities are most valuable to them and which ones could be dropped. This would require assigning "fair shares" of pollution and resource use allowances to individuals, which is, as argued above, not realistic as a catch-all policy approach.

The only feasible alternative, in order to avoid being arbitrary, therefore seems to be a careful assessment of which areas of consumption contribute most to accumulative environmental harms. Such assessments exist: there is a broad consensus among researchers that most adverse environmental impacts from consumption in industrialised countries are produced within the areas of mobility (particularly automobiles and air transport), food consumption (meat, dairy, and other food items), housing or buildings, and the use of electric appliances in homes (including their life-cycle impacts). In total, these areas amount to approximately 70% to 80% of the environmental impacts from consumption in rich countries (Tukker et al. 2006, 2010, cf. also EEA 2010: 15).

These broad categories can be further broken down to gather more specific information on concrete (groups of) goods and services. With regard to these, it can be evaluated which technical solutions (that do not involve significant impacts on individual freedom) are available and in how far they can successfully remedy the adverse impacts emanating from a consumption area. Where technical solutions are not effective enough, too expensive, or involve too strong side effects, sufficiency policies would have to be applied in addition.

Besides such a technical assessment, it seems unavoidable to make judgements concerning the importance and avoidability of certain consumption practices in order to meet human needs. The "gas-guzzler" might then be among the victims of such necessary prioritisations. Likewise, in cases of obvious and perceivable resource scarcity, for instance when there is a local water shortage, measures that prohibit less important use of water, such as filling swimming pools, may be deemed appropriate.

A clear advantage of policy solutions that focus on consumption practices or goods is that they address the options and underlying structures based on which individuals make their choices. This means that individuals are not left alone with the requirement to reduce, say, their greenhouse gas emissions to a certain level, but that infrastructure and other conditions can be designed in a way that supports them in doing so (cf. section 4.1.1).

To sum up, it seems that the ideal solutions with which to reduce individual consumption proposed by economic theory (internalise external costs) or justice theory (distribute allowances to pollute and use resources according to a fair distribution principle) are unrealistic in practice as "one-size-fits-all" policies. Of course, these instruments can be applied but not as standalone solutions, which would produce too many undesirable side effects. Supplementing them with a piecemeal approach that targets specific acts of consumption is not ideal in terms of efficiency and non-arbitrariness, but often superior with regard to feasibility. Examples of all three solutions will be discussed in section 5.2.

4.4 Potential toe-holds for criticism

Finally in this chapter, which has presented my approach to justifying sufficiency as policy based on the harm principle, I would like to consider some perspectives from which this approach might be criticised. In the following, I will try to anticipate some of these and give according answers. Criticism may come in particular from those who still see a threat to the neutrality principle, from those who fear too strong interference in individual liberty regardless of the potential harms involved, from those who see the harm principle as too weak a basis and would rather give up neutrality, as well as from those who do not find it realistic for sufficiency policy to be realised at all in a democracy.

4.4.1 "Sufficiency policies violate the neutrality principle"

Even though the preceding analysis is based on the harm principle, which is in principle accepted by liberals as a legitimate reason for interference in individual liberty, interference in consumption activities could still be seen as a hidden attempt by environmentalists with an anti-materialist attitude to impose their view of the good life on others (cf. Heath 2005: 210). This scenario is actually not so unlikely, given the anti-consumerist attitudes of many environmentalists who also tend to base their arguments on the good life.[223] My intention here is not to argue against such perfectionist accounts, which I value—if they are elaborated well—as interesting and enriching, but rather to distinguish my arguments from them in order not to be "accused" of perfectionism. While the ideal of neutrality towards the good is controversial, as explained before, I have stated that I accept it as it is a basic element of "mainstream" liberalism (cf. section 2.1.3).

There appear to be two lines of discussion within the literature on this subject which ask whether neutrality and environmentalism are compatible. One focuses on policies that seek to protect specific nature sites and species, for which it is rather difficult to argue that the principle of avoiding harm to human beings requires them[224]. However, such policies are often characterised by a rather low degree of interference in personal freedom. The question is thus whether such policies should still be adopted, even if not directly required by the harm principle.

The second line of discussion is concerned with the way of life in industrialised countries and the question of whether liberal (and therefore, in mainstream terms, by definition neutralist) policy is allowed to influence consumption patterns for the purpose of environmental protection. Such policies, contrarily to the ones that aim at nature protection, seem, as argued above, to be required to some extent by the harm principle, but they are also characterised by potentially significant interference in personal ideas of the good. The question, here, is how far such environmental policies can go in order to be still in line with neutrality.[225]

This is, I think, an important differentiation which is not considered sufficiently by some analyses of the compatibility of environmental policies and the neutrality principle. Sagoff (1988) and, based on Sagoff's arguments, also Meyer (2011) claim that discussing this compatibility is a "category mistake", arguing that environmental policy does not interfere

---

223 Some consumption researchers are very critical of such attitudes, accusing the anti-consumerists of their disregard of the important social functions of consumption (cf. e.g. Miller 2012).

224 Cf. the section on biodiversity loss in chapter 1. The loss of biodiversity in general, as recent scientific findings indicate, seems to indeed have harmful consequences for human beings. However, it would be more difficult to claim this for the loss of a particular species or nature site.

225 The differentiation between these two lines of discussion appears to be broadly in line with the distinction between "basic" versus "non-basic" environmental goods (cf. Miller 1999 a, 2004, Humphrey 2003, Hannis 2005).

with individual conceptions of the good. Both therefore right away reject the usefulness of the discussion. According to Sagoff (1988: 165 f.),

> *"Environmental decisions, by and large, have to do with what goes on out of doors not indoors; they concern the character and quality of the public household, not the private home. Environmental policies, in general, restrict what corporations and municipalities may do with their investments and effluents—not what individuals may do with their lovers, co-worshipers, or friends. Thus the content of environmental policy rarely becomes relevant to the kind of neutrality essential to liberalism."*

Sagoff and also Meyer obviously mainly have examples of nature protection policies in mind, and do not take into account the debate on influencing lifestyles. Stating that environmental policies do "by and large" not affect individuals in their daily lives is an oversimplification. Today we see plenty of such policies that regulate the energy efficiency of houses which may affect the aesthetic appearance of a building, of appliances that individuals can buy, or that determine which standards their cars have to meet; even if such efficiency policies only affect individuals to a limited degree, as I have argued above (section 4.3.2), potential environmental policies in the sense of sufficiency could affect individuals' life plans much more. These policies do, of course, not have an impact on the religious or sexual activities of individuals, which Sagoff considers to be at the core of what the neutralist state should not influence. However, they still can have an impact on personal conceptions of the good (Coglianese 1998: 42) if, for instance, driving around in a highly polluting vintage car is seen as part of this good.

How deep this impact is and if it can be justified by the harm principle needs to be assessed on a case-by-case basis. However, some general considerations can still be made. In the following, I will do this separately for the two strands of discussion, starting with the second one—regulation of consumption patterns—as this is the one which is mainly of interest for this study.

As argued before, it is in principle possible to establish environmental policies that will have an impact on individual lifestyles without violation of the liberal neutrality principle if these policies are needed to curb accumulative environmental harm. The discussion on accumulative harms often deals with serious potential harms to others, including risks to life or serious health risks. Therefore, policies to stop such harms are clearly not perfectionist, since serious (potential) harms to others provide for a neutral justification. If one defines the neutrality principle in such a way that it requires only neutrality of the *justification* for a policy (cf. section 2.1.3), then it is satisfied in such cases. However, what if such a policy caused relevant non-neutral *effects*? If the neutrality principle is also conceived as encompassing the neutrality of effects, then policies affecting lifestyles would be more problematic.

Coglianese (1998: 44 ff.) adopts such a view, arguing that more care needs to be taken in assessing the potential non-neutral effects of environmental policies, even when they can be justified on neutral grounds. He does not endorse a very strict view, which would prohibit all policies that produce non-neutral effects (or require the compensation of all such effects, cf. section 2.1.3) and would be unrealistic. Instead, he adopts a gradual approach, stating that the "more serious the non-neutral effects are, the more substantial the justification must be" (Coglianese 1998: 47). "More substantial" justifications in this sense are those which aim at preventing greater harm, have a clearer causal chain, or are linked to a higher likelihood that the interference will be effective. Coglianese is particularly worried about speculative claims of potential environmental harm based on extended causal chains. He objects that if governments only have to claim that "there may be some adverse effect somewhere, at some time, on someone" in order to satisfy the harm principle, then the principle would become useless (1998: 49). He gives, among others, the example of car driving (1998: 50):

> *"A thin causal claim about harm may be sufficient for regulation that affects individuals negligibly, such as by requiring them to purchase automobiles that meet reasonable emissions standards. The same thin causal claim would probably not be sufficient, however, for prohibiting individuals from driving cars altogether."*

Indeed, imperfect knowledge about causalities and the probability with which environmental harm will actually occur, as well as the difficulties with the interpretation of the term "harm" impede a clear-cut scheme to identify cases where environmental policies can be legitimately based on the harm principle. Therefore, Coglianese's quite pragmatic proposal that more serious non-neutral effects require more substantial justifications is quite valuable and more useful than approaches to neutrality which assume that it is possible in practice to realise strict state neutrality that never intervenes in citizens' conceptions of the good.

However, Coglianese omits two important points, as Barry notes (2001: 68 f.). First, not taking action in the sense of environmental policies can also have non-neutral effects (cf. also Horstkötter 2004: 162), which, for my discussion here, is already implied in the basic assumption that there is (or may be) an instance of harm. If no action was taken to curb greenhouse gas emissions, it is possible that millions of climate refugees would have to leave their homes, which would drastically affect their lives, which is certainly different from what these individuals consider to be their vision of the good life.

Second, Coglianese argues that a firm causal relationship between environmental effects and harm to human beings needs to be established before the state is allowed to intervene, even in cases where environmental problems are "potentially catastrophic" (1998: 59)[226]. As opposed to that and in the name of the precautionary principle, environmentalists demand that this burden of proof is reversed, that is, that those in favour of an activity suspected of bringing about harm should prove its harmlessness (Barry 2001: 69). However, the application of the precautionary principle is also seen as problematic by some (cf. Box 2, section 2.3.4), and probably even the fieriest environmentalists would agree that not every suspected fragile link of some activity with some potential harm should lead to this activity being banned. This is, again, a matter of degree which is case-specific, involving considerations of the credibility of the available scientific evidence, the immediacy of finding a solution, the costs (or "non-neutral effects"), and the risks and uncertainties linked to each case.

With regard to the second strand of discussion identified above, it is much more difficult to establish a connection between the loss of a species or a specific nature site, for instance a spectacular landscape such as the Grand Canyon, and (potential) human harm (Coglianese 1998: 48). For the loss of biodiversity in general, one can of course argue that beyond a certain point there would be harmful consequences, as is suggested in the idea of a planetary boundary to biodiversity loss (cf. chapter 1 and section 2.3.4). To take a (unrealistically) drastic example, an entire category of renewable resources (or just a relevant share of that category), such as fish, might be wiped out, which would deprive some parts of mankind of their livelihood (cf. the discussion in section 4.3.4). Also, one could say that if there were no natural sites left as recreational areas, people's health would be harmed because they would have to stay in a stressful urban environment all the time. However, restricting nature conservation policies to such extreme cases would not at all satisfy environmentalists' demands.

226 Cf. section 2.3.4, Box 2 on potentially catastrophic environmental harm.

As already argued above, with regard to this second strand of discussion, not taking action in the sense of environmental policy can also have non-neutral effects. If, for instance, the only nearby forest is not protected but sacrificed to the timber industry, citizens are prevented from enjoying the forest in their leisure time, which might have been an important component of their idea of the good life. The other way round, this means that environmental policy can also increase, not only limit, the availability of conceptions of the good from which individuals can choose (cf. de-Shalit 2000: 83 f.). However, as de-Shalit notes, this way of arguing can also be misleading: the aim of keeping the choice among the variety of ideas of the good as extensive as possible alone would also mean that things should be conserved that are commonly considered bad (e.g. nuclear weapons). He therefore concludes that an argument in favour of the conservation of an object has to be based on the claim that the object under concern is good as such (de-Shalit 2000: 84).

Still, environmental policy measures that cannot be justified as required to avoid harm do not necessarily violate the neutrality principle. Liberal theorists offer at least three ways to allow for such policies to be adopted and still be in line with the principle:

- If such natural assets are included within a conception of distributive justice, such a conception can require them to be distributed adequately together with other goods or resources within and between generations (Dworkin 1978: 141, Coglianese 1998: 46). It should, however, be noted that there is some discussion about which goods[227] can be included in a justice conception (cf. Miller 1999 a, 2004, Bell 2002, Humphrey 2003, Hannis 2005).
- If the degree of interference is low and the provision of some environmental good would therefore "not pose substantive issues of justice", democratic reasoning could be applied, and if there is sufficient agreement this would lead to provision of the environmental good (Miller

227 An iconic case from this discussion is the snail darter, a small fish in the US which was threatened with extinction in the 1970 s by plans to build a dam in its habitat that would have blocked its migration route. Several theorists refer to the snail darter in their discussion of distributive justice and the environment (e.g. Coglianese 1998: 49, Miller 1999 a: 160, 2004: 141, Hannis 2005).

1999 a: 171 f., Bell 2002)[228]. However, doubts have also been expressed about whether such a "neutral" procedure will lead to any substantial level of environmental protection (cf. Hannis 2005: 586 ff., see also Bell 2002: 721).

- If animals or other elements of the non-human environment were included in the moral community, that is, if a physiocentric perspective was adopted, their rights would have to be protected independently of human interests (cf. section 2.3.2).

I will not discuss these approaches any further here, as they go beyond my own very basic account developed in previous chapters.

Concluding on this point, I see my judgement reinforced that policies that aim to influence consumption levels in order to decrease their environmental impacts can, in principle, be justified on a neutral basis by the harm principle. However, if such policies are expected to have significant non-neutral effects that cannot be avoided, the justification should, in line with the principle of commensurability, be rather substantial (cf. section 4.3.5). Risks of severe environmental harms, such as climate change, will usually be appropriate as substantial justifications. Policies that aim at nature protection for which a case of (potential) harm cannot be established need to be based on different reasoning.

4.4.2 "What is at stake is too important"

Even if in line with the harm and the neutrality principle, and hence with mainstream liberalism, one could still say that political interference in individual consumption levels poses a threat to foundational values prevalent in the Western civilisation, and thus to something that is simply too important to give up. In this sense, Schramme remarked in an article on consumer choices (2011: 352):

> *"[...] even when our life is in danger, and, indeed, when we endanger other people's lives by our choices and actions, we still tend to think that it might be worth it. Thus, we smoke, use cars, invent new materials that sometimes prove to be unhealthy, etc. We do all these things, although we often kill innocent people by doing them, because we tend to think that our individual liberty is*

228 However, see also Miller's (1999 a) rejection of providing any environmental goods going beyond this, which would pose substantial issues of justice, and Bell's (2002) related discussion of this.

more valuable than living under conditions that reduce all risks to life and limb. But this is essentially what the environmental crisis is all about: posing a threat to our life and limb. Whether we therefore have license to criticise and ban everything that might be involved in that threat remains to be demonstrated, in my opinion. [...] Perhaps safeguarding a healthy environment will ask too much of us. I, for one, would not give up individual liberties, of which consumer choice is one example, light-heartedly, even if the stakes are high."

However, this view appears to be incoherent with the liberal perspective, to which the harm principle is a very basic element: we may endanger our own lives, but we are not allowed to endanger the lives and health of others as long as they have not consented to this and as there are no equally substantial reasons for doing so. The freedom to consume as much and whatever one likes does, in my view, not meet this latter condition of a substantial justification for imposing harm on others.

From a liberal point of view, we are simply not in a position to decide that "it might be worth it" on behalf of those who cannot influence the decision and will suffer the adverse effects of rich countries' current consumption patterns, such as future generations and people in poor areas of the world who are affected, for instance, by global climate change.

In addition, influencing consumer choices does not necessarily mean banning things or even restricting liberty. It may also mean abolishing protectionist and environmentally unsound subsidies to make it easier for individuals to choose less harmful alternatives, or to provide them with information about the consequences of certain consumption patterns (see chapter 5).

4.4.3 "Without any foundation in an account of the good life sufficiency policy remains ineffective"

From a perspective quite contrary to those outlined before, say, one devoted to "bolder" environmentalism, one could criticise the approach outlined in this study as too defensive and unambitious. One could say that without one grounding it in a perfectionist account, that is, an account of the good life, sufficiency policy would remain ineffective and powerless, and the arguments raised in relation to harmful pollution and high resource use stand on rather shaky ground (e.g. Sagoff 1997, Skidelsky/Skidelsky 2012).[229]

Using harm as a justification serves only to restrict some consumption activities, as the causal link between consumption and harm might not always be strong enough (cf. the discussion in section 4.3.4). In particular, as stated above, protection of valuable landscapes and ecosystems or conservation of certain species may be very difficult or even impossible under a scheme merely based on the harm principle.

Hannis, for instance, argues that "only the most anaemic and uncontroversial environmental regulation could be justified" on the basis of an account of environmental goods that are required to satisfy basic human needs (2005: 591). He therefore states that effective environmental policies are simply not compatible with the principle of state neutrality towards conceptions of the good (cf. also de-Shalit 2000: 83). In his view, environmental policies need to be interventionist and have to be based on normative principles, especially as regards levels of production and consumption, which would be best "defined and justified by reference to some substantive conception of human flourishing" (Hannis 2005: 577).[230]

Indeed, I have proposed a kind of "minimal" approach to sufficiency policy that relies on several least controversial positions: anthropocentrism, which assigns nature with no more than an instrumental function to serve the human good; a justice conception focusing on meeting basic needs; and allowing for state interference particularly on the basis of avoiding harm to others.

However, I am not proposing basing *all* environmental policy on the harm principle, even though I have argued that actually many environmental regulations *can* be justified this way. Yet, sufficiency policy has the potential to interfere substantially with individual freedom, therefore the justification for such a policy must be especially robust. Other environmental policies, such as those aimed at nature protection which interfere less with individual liberty, can more easily be based on less strong justifications (cf. section 4.4.1).

229 Cf. the following quotation from Robert and Edward Skidelsky's book *How much is enough?* (2012: 132): "It is as faith, not science, that environmentalism stands or falls." They state this in relation to harm principle-like arguments and argue for a perfectionist account instead.

230 Cf. also the more nuanced view of Stephens (2001 a, 2001 b).

One could also object that using harm as a justification corresponds to a negative perspective on environmental protection and is thus unattractive, recalling apocalyptic scenarios promoted by the environmental movement in the 1970 s. In this sense, limits on consumption are seen as unavoidable, even though nobody wants them. A good life approach would instead yield the chance to provide a more positive, even attractive perspective for consumers. This is, for instance, promoted by Kate Soper's "alternative hedonism" (e.g. Soper 2007, 2008 a, 2008 b), which highlights motives for adopting less materialistic consumption patterns that lie in the positive effects for consumers themselves, such as enjoyment from walking or cycling and other pleasures that are lost in current work-dominated and high-speed lifestyles. In addition, these motives can be or are in practice often complemented with more altruistic ones such as a concern for the Earth's carrying capacity and social justice (Soper 2007: 211).

However, Soper's account of the good life is not, or at least not explicitly, conceived as a stand-alone justification of policies that aim to bring about lifestyles in the sense of "alternative hedonism". Soper rather vaguely brings up notions such as "demand of social justice" and "global ecological security", which can be interpreted as (additional) justifications (2007: 223). Instead, she primarily seeks to highlight the importance that "alternative hedonist" attitudes have in providing sufficient support for certain policies and that these should be encouraged (Soper 2008 b: 198 f.):

> *"[...] in the absence of a revised conception among the already more affluent communities of their own needs, interests and pleasures, little pressure is likely to be applied to secure the necessary policy changes. A changed perspective of the 'good life' (an 'alternative hedonism') is in this sense an essential stimulant to any possible future curbing of consumerist culture and its more negative social and environmental impacts."*

In this sense, arguments related to the implications for the good life can be applied in political communication, but it would be quite difficult to establish the main foundation for sufficiency policy based solely on such arguments.

4.4.4 "In a democracy, sufficiency will never be adopted as policy"

As indicated in section 4.1.2, there are doubts about whether sufficiency, at least explicitly, will ever be adopted as policy within democracies, as this would require that a majority is in favour of such a policy (cf. Huber 1995 a: 129 f., 1999: 12, Stengel 2011: 228, Heyen et al. 2013: 18). Some have claimed that before this happens, the "environmental crisis" would have to become so compelling that there is no other choice left (cf. section 4.1.2). This scepticism stands in the tradition of earlier environmentalists who doubted the capacity of democratic states to adopt meaningful environmental policies (cf. section 2.2.4).

However, this is obviously essentially an empirical question. Two empirical facts do provide some support that it is not impossible that sufficiency may be adopted as policy more systematically in the future: first, there were not many environmental policies in most countries before the 1970 s, so it is a field which is rather new and only became increasingly accepted over time. Today, as a result of this cultural change (Dryzek/Dunleavy 2009: 219), environmental regulations are a common part of the law in countries worldwide[231], and in many industrialised countries established political parties can hardly afford to do without an environmental programme. Thus, it is not unreasonable to expect that this trend will continue and stricter environmental policies, including those that have the effect of limiting consumption, will be possible in the future. Second, there is some indication that young people in industrialised countries adopt lifestyles that rely less on material assets and use of environmental resources compared to earlier generations (cf. Soper 2008 b: 201). For instance, access to cars and the availability of driving licences have tended to decrease (Kuhnimhof et al. 2012), the number of those who do not eat meat or even do not consume any animal products has increased steadily (Leahy et al. 2010), and in some milieus resource-light lifestyles have become trendy over the past decade. It can thus be concluded that the share of people who are likely to accept sufficiency as policy is increasing, and, reversely, with their attitudes and behaviour help to prepare the ground for such a policy in the sense of the interdependence between individual agency and societal structures (cf. section 4.1.3).

Research on environmental policies has also shown that situative conditions play a role in the potential of realising a certain policy (cf. Jänicke 2003: 79 f.). "Opportunity windows" can significantly enlarge the chances of a policy being realised successfully. Further, in deliberative democra-

231 See, for instance, the (incomplete, but long) list on https://en.wikipedia.org/wiki/List_of_environmental_laws_by_country (last accessed November 2015).

cies that rely on an exchange of arguments, it is up to the environmentalists to develop arguments that are accessible to their fellow citizens and persuade enough of them to change their preferences and have an impact on policies (Saward 1993: 77, de-Shalit 1997: 181, Bell 2002: 721; cf. Gutmann/Thompson 2004: 4, 20). It is my intention in this study to provide such arguments that are accessible to as many people as possible by using an approach about which agreement should, in principle, be high.

Finally, in order to ease the adoption of sufficiency as policy within liberal democracies, it makes sense to start with policies that also yield gains for individuals and to implement them incrementally (Linz 2017), even if this should not undermine other criteria of policy selection, such as effectiveness. Success criteria and specific policy options will be dealt with further in chapter 5.

4.5 Interim summary

This chapter distinguished between sufficiency as an individual and as a political approach. It was argued that, while the acts of individuals or individual households in the sense of sufficiency are important, for instance as examples to others or to manifest a trend, respective action should also be taken on a political level, as sufficiency as an individual strategy is not effective enough to bring about the changes in consumption patterns needed to reduce macroeconomic consumption to a sustainable level. Mainly the following reasons for this limited effectiveness were identified: the effects are limited in quantitative terms if only a minority changes its behaviour; the knowledge of individuals about the complex environmental consequences of their consumption patterns is limited; there can be macroeconomic rebound effects; individual action can be used as an excuse not to take collective action; individual behaviour is embedded into and constrained by the existing systems of provision, such as the infrastructure, economic circumstances, and technical systems. The "political" was outlined rather narrowly as a notion that refers to actions by the state on its various levels, resulting in decisions that are collectively binding. It was also mentioned that the individual and the political levels are interdependent. Not only do the surrounding structures constrain and enable individual action, but, reversely, individuals also influence and shape these structures through their adoption and continued performance of daily rountines.

The literature from the sustainability discourse already features a number of publications dealing with sufficiency as policy, but these mostly take a rather practical perspective and lack a detailed analysis of the justification for and objections to such policies. In search of such justification, the liberal principle that society may intervene when there is an instance or a risk of harm to others ("harm principle") was discussed in detail. It was argued that a case can be made for the harm principle as an adequate justification for most policies that aim to protect the environment, also when the harm is the outcome of many uncoordinated individual acts ("accumulative harm"). Even if the harm principle enjoys widespread acceptance, there is much controversy about the crucial notion of "harm", and there are doubts about the possibility of finding a way to define harm objectively. In the case dealt with here—environmental change that has adverse effects, for instance on human health—, however, it will often be clear whether there is an instance of harm or not. Discussions about environmental harms are more likely to concern the threshold of harm rather than the question of whether something should generally be characterised as harm or risk of harm. Cases of accumulative harm are caused by many small contributions from individuals (or other parties), which are, in themselves, harmless, but in higher numbers they accumulate in the environment and approach a threshold of harm over time. It is comparably difficult to find appropriate ways to deal with accumulative harm, since the causal chain for the harm is not exactly traceable and such harms will often materialise only with a certain probability. However, to argue that the costs of avoiding such harms are too high does not hold as a general excuse given that there are potentially much more significant costs on the side of those harmed. An instance or risk of harm does not create an automatism that a regulation must be set in place, but does open the discussion about possible measures to deal with the harm. As soon as it is sufficiently clear that the parallel actions of individuals lead to an accumulative problem that brings about harm, this gives the community a reason to organise itself in such a way that the activities contributing to the harm are reduced so that the threshold of harm is not reached.

The harm principle therefore also holds as a justification for policies that aim to limit environmentally harmful consumption. However, it was pointed out that finding ways to stop accumulative harms due to consumption levels is particularly difficult, because compared to technical measures, sufficiency policies tend to entail deeper infringements of individual liberty, as they imply quantitative restrictions and require behaviour changes, while technical policies "only" aim at improving products and processes. The harm principle itself does not provide an answer to the question of how society should intervene in cases of harm to others. In particular, as consumption refers to a wide and inhomogenuous set of activities that are often valuable to those who perform them and may also be beneficial to society, it should be considered how deeply a measure interferes with individual liberty. In line with the principle of commensurability, the reasons for adopting very compulsory measures must be more significant than those for less compulsory ones, and there may not be any equally effective and less intrusive solutions available. Prohibiting an activity will often not be an adequate solution. Besides the option of curbing consumption levels by means of focusing on specific, environmentally unsound activities, two other solutions were considered: focusing on pollutants or scarce resources, and focusing on individuals. All three have their advantages and disadvantages, and—unlike theorists sometimes claim—none of them represents an ideal "catch-all" solution. In addition, it was pointed out that justifying a policy that aims to limit the unsustainable exploitation of resources on the basis of the harm principle is comparably more difficult than a policy that aims to curb the pollution of environmental sinks. This is due to the fact that accumulative harms that may, among other things, be driven by the high resource consumption of some people are even more indirect than harms caused by pollution, given that there are many interacting mechanisms relevant to globalised markets. Still, this does not exclude the possibility of (or requirement for) such policies per se.

Finally, this chapter anticipated a number of critical perspectives on my approach to sufficiency as policy and gave according answers. Among these, potential fears about violations of the neutrality principle were considered in detail, and it was again concluded that these are, if a clear case for the harm principle can be made, unjustified. Further, it might still be said that interference in consumption levels puts things at risk that are too important to give up and that such interference should thus be refrained from; this reasoning would, however, not be in line with the harm principle. From yet a different angle, it was considered that the harm principle might be seen as too defensive a basis for meaningful environmental protection measures, with the result that neutrality should rather be given up. However, as my intention is to provide a robust justification that is as broadly acceptable as possible, I do not consider perfectionism an appropriate basis. Still, the more positive aspects of a perfectionist account related to the pleasures of lifestyles in line with sufficiency can be used in communication along with sufficiency policies. Finally, it might seem unrealistic that sufficiency policy will be realised at all in a democracy. With regard to this, it can only be stated that this is an empirical question, and that there are certain tendencies in wealthy countries that indicate an increasing interest in lower consumption lifestyles, which can pave the way for according policies. Robust arguments on the side of those advocating sufficiency as policy will be helpful in deliberation processes to convince a sufficienctly great number of people.

As a result, it can be said that adopting sufficiency as policy, that is, recalling the definition of sufficiency adopted in chapter 3, policies that attempt to reduce the demand or use of goods and services with high external environmental effects in order to achieve sustainable per-capita consumption levels is both necessary and possible within liberal-democratic societies. The problem, therefore, now consists less in the doubts about whether this would, in principle, be compatible with liberalism, but rather in the question of *how* this might be done in order to be effective and feasible, and affect liberty as little as possible. The next chapter will deal with such practical questions and specific policy instruments.

5. Sufficiency as policy: potential solutions

> *"There are several reasons why our consumption standards have not been called in question in the course of the conservation discussion over the last fifty years. [...] it will be said that there is nothing that can be done about consumption. This of course is nonsense. There is a wide range of instruments of social control. Taxation; specific prohibitions on wasteful products, uses, or practices; educational and other hortatory efforts; subsidies to encourage consumption of cheaper and more plentiful substitutes are all available. [...] Since consumption could not be discussed without raising the question of an increased role for the state, it was not discussed."* (Galbraith 1958 b: 95 ff.)

5.1 Overview of policy instruments

Indeed, as economist John Kenneth Galbraith already noted in 1958, there are in principle many policy instruments available that could be applied to interfere in levels of consumption. These, however, have only been used to a limited extent, as concerns regarding consumer choice and freedom as well as conflicts with the paradigm of economic growth have prevailed. Instruments of environmental policy, according to political scientist Martin Jänicke, include all options for action available to environmental actors in order to realise environmental aims (2003: 100). This also seems to include rather informal activities, such as launching a public debate, or activities that generate knowledge, such as supporting environmental research and experimental settings[232]. I use a more narrow understanding in the following, which focuses on mainly legislative instruments that aim to reach a large share of consumers.

This section gives an overview of policy categories and more specifically concrete proposals for sufficiency policies that could be implemented in industrialised countries, that is, policies that aim to decrease demand or use of goods and services with high environmental impacts in order to achieve per-capita consumption levels that ensure emissions and resource use to stay within the environmental carrying capacity.

232 Note that, despite the broad definition, such forms of action are not part of Jänicke's categorisation, which is cited below.

5.1.1 Categories and assessment criteria

A popular categorisation of environmental policy instruments has been proposed by Jänicke (2003; see Table 8). It sorts policies according to their "degree of the state's determination of individual behaviour", or, in other words, to the degree of interference or even coercion involved (cf. section 4.3.5). Jänicke distinguishes between five basic categories, listed here by decreasing degree of interference: regulatory law, planning, economic instruments, cooperation, and information[233]. In principle, all these instruments could be used to implement sufficiency as policy, and they contain what the literature on sufficiency often refers to as constraining approaches at one end as well as enabling strategies at the other (cf. section 4.1.2).

Table 8: Categorisation of environmental policy instruments[234]

Group of instruments	Instruments (examples)	Degree of the state's determination of individual behaviour
Regulatory law	– orders and bans – licensing – setting maximum permissible values – product and process standards – environmental penal law	high
Planning	– regional development plans – land use plans – urban and transportation planning – air quality control plans – waste management plans – water management plans	high/medium

233 Note that the categories of policy approaches that I have developed in chapter 4 (see section 4.3.6) cannot be expressed in terms of Jänicke's categories. Rather, they are located across the latter. In the second part of this chapter, I will come back to my own categories by discussing an example from each of them (see section 5.2).

Group of instruments	Instruments (examples)	Degree of the state's determination of individual behaviour
Economic instruments	public revenue: – environmental taxation – environmental charges – fees public spending: – tax reliefs – subsidies – green public procurement others: – pollution certificate trading – user advantages – environmental liability	medium
Cooperation	– negotiations – forming networks – sector agreements – voluntary agreements	medium/low
Information	– information campaigns – environmental labels – environmental education	low

The choice of the most appropriate environmental policy instrument to solve a problem depends on the circumstances of each particular case, and often a combination of instruments is applied. Common criteria by which to judge the appropriateness of environmental policy instruments include effectiveness, (cost) efficiency[235], the likelihood of implementation deficits, and political feasibility (Röckseisen 1998).[236] In the context here, a further criterion is of particular relevance, namely the impact on individual liberty. These five criteria will be used later on to assess different poli-

234 Translated and slightly amended from Jänicke (2003: 101).

235 As explained in section 3.1.3, efficiency aims at the optimisation of a cost-benefit ratio. While efficiency as a sustainability strategy as outlined in section 3.1.3 focuses on the optimisation of the ratio between environmental burdens and overall benefits, as a criterion for the analysis of policy instruments it focuses more on other, for instance, financial costs.

236 Röckseisen also mentions legal conformity, which is not so much in focus here, as it depends on the use of a specific instrument within a certain jurisdiction.

cy proposals and are closely linked to what has been laid out as the framework for analysis in chapter 2. Effectiveness describes the contribution of a policy instrument towards the aim of achieving sustainability (section 2.3). Political feasibility is determined by the chances of an instrument being realised, which tend to be high if it fits well into existing structures (Röckseisen 1998: 177) and which depend on public acceptance of an instrument. This criterion therefore relates to the democratic need to get the support of a majority after careful deliberation, or at least not to meet with resistance from the majority (section 2.2, see also section 4.4.4). The impact on individual liberty is, of course, a central concern of liberalism (section 2.1, see also sections 4.2, 4.3). As regards the cost efficiency criterion, to take this into account is simply rational in view of the limited financial and other resources available for government tasks. Finally, implementation deficits occur when the agent who is responsible for realising or complying with a policy is not able or does not want to fully conform to this task. They are closely linked to both effectiveness and efficiency (Röckseisen 1998: 175) as they have the potential to undermine achieving these criteria.

As a side remark, it should be noted that Jänicke himself (e.g. 1996: 11 f., 2003: 108 ff.) and others have observed and criticised a certain obsession among environmental policy analysts with policy instruments. Based on empirical evidence, it was found that determining overall targets as well as the communication processes about certain problematic environmental developments and related behaviour in the context of a potential policy intervention are much more important than the choice of a concrete policy instrument. This criticism, however, is directed in particular towards those who tend to praise the advantages of a certain category of instruments ("only regulation is effective", or "only economic instruments are effective *and* efficient", etc.), while it still generally seems to make sense to discuss the appropriateness of a certain policy instrument to solve a certain problem.

5.1.2 Policy proposals in the literature

A number of authors have provided specific proposals of what a sufficiency policy or a policy to curtail consumption could comprise (cf. Røpke 1999: 417 f., Sanne 2002: 284 ff., Darby 2007, Linz/Kristof 2007, Ott/Döring 2007: 62 f., Linz/Scherhorn 2011, Linz 2012 a, 2015, Brischke et al. 2011, Brischke/Spengler 2011, Costanza et al. 2012: 54 f., Seidl/Zahrnt 2012, Voget-Kleschin 2012: 231 ff., Heyen et al. 2013: 13 ff., Schneidewind/Zahrnt 2014)[237]. Some of these proposals deal with the kind of policies that may be applied (or should not be applied any longer), ranging, in line with the above categorisation, from less to more behaviour-determining policies, from information campaigns to "hard" regulatory law. Others give lists of examples of concrete instruments that could be established to start with, considering which measures would receive mostly broad acceptance among the public and help mainstream the idea of sufficiency (cf. Seidl/Zahrnt 2012)[238]. Overall, literature sources indicate that in order to deal with consumption levels, a variety of measures is needed, which might result in a rather complex policy pattern (e.g. Røpke 1999: 417 f., Jackson 2007: 262).

Most of the ideas are not completely new but represent policies that are, in principle, already implemented in liberal democracies. For instance, environmental taxes or city infrastructure reforms are common practice, but in the sense of a sufficiency policy they would have to be implemented much more consistently in order to influence consumer practice effectively in terms of consuming *less*. Even if the proposals all[239] match my (rather broad) definition of sufficiency as policy (cf. section 4.5), there are a number of measures that aim, in the first place, at the production side rather than the consumption side[240], for instance design requirements on product lifetime or absolute limits on car emissions. Nevertheless, there would also be relevant consequences on consumption, such as perceivable price increases or as regards the availability of certain goods (e.g. cheap goods with a short lifetime, cars of a certain size/weight).

Most proposals relate to the consumption fields that are environmentally most relevant (cf. section 4.3.6)—the mobility/transport sector, food, housing/buildings, and electrical appliances or other products—but many are also cross-cutting and are not directed at any specific field. They can

237 Specifically on the issue of "nudging" (see below), sources apart from the sufficiency discussion were added (Smeddinck 2011, Croson/Treich 2014, Sunstein/Reisch 2014, Reisch/Sandrini 2015), as this has not (yet) been dealt with comprehensively in the literature on sufficiency.

238 See also the success criteria for sufficiency policy listed in Linz (2017).

239 Except for those that have been excluded here and are not listed in Table 9 for the reasons explained further below.

240 Cf. Table 4 for the distinction between sufficiency and technical strategies in terms of consumption versus production.

by and large be sorted according to Jänicke's categories of policy instruments, as introduced in the last section. As an amendment, an additional policy category was created, which comprises the idea of "nudging" people into environmentally sound decisions (cf. section 5.1.3)—an idea that was raised and has has only gained momentum in recent years and is therefore not considered in Jänicke's categorisation[241]. It is here placed in between economic instruments and cooperation, according to its estimated degree of state determination of individual behaviour. Ordered this way, Table 9 presents the policy proposals from the sources cited above.

Further, I have excluded measures that would be based on a specific vision of the good life rather than on environmental harm (e.g. "less concrete and monotony in urban planning", see Schneidewind/Zahrnt 2014: 85), as well as proposals that are not policy instruments as defined above (section 5.1). In particular, several literature sources mention sufficiency "policies", which are not policy instruments in the narrower sense, but rather point to more far-reaching socio-economic developments that would require a fundamental change in values and entire programmes for far-reaching political reform. Examples include "abandon economic growth policy", "reduce work time to support self-subsistence and non-consumptive leisure activities and to break the work-and-spend cycle", "decrease social inequality to reduce luxury and 'conspicuous' consumption", and "curb the momentum of sociotechnical systems towards ever-increasing standards". These could indeed have significant effects in terms of "mainstreaming" sufficiency. However, as they are not concrete policies that could be adopted and implemented by the state, I exclude them from the following analysis. Of course, they raise the question of whether it is at all possible to mainstream sufficiency with a narrow focus on policy instruments alone. More fundamental changes are certainly necessary and are a prerequisite for realising a sufficiency policy. It is, however, my aim here to discuss only possible measures taken by the state (cf. the "political" in the sense defined at the beginning of section 4.1), while I do not deny the importance of other actors and developments.

241 Cf. also Smeddinck (2011: 380 ff.) on this, who tries to describe nudging by means of the traditional instrument categories.

Table 9: Overview of sufficiency policies suggested in the literature[242]

Category	Sufficiency policy instruments
Regulatory law	Mobility: – speed limits – prohibit driving private cars on certain days (e.g. Sundays) – limit the absolute energy use or carbon dioxide emissions of cars Food: – establish stricter environmental standards for food production, prohibit factory farming (which will at the same time increase prices) Housing/buildings: – upper limits for the energy use of buildings Appliances/products: – progressive efficiency standards that are stricter for bigger or more powerful energy-using products – upper limits for the energy use of products – extend minimum product warranty to discourage short product lifetimes – prevent early "obsolescence"[243] with design requirements for long lifetimes, repairability, reusability – ban certain goods (e.g. outdoor heaters, plastic bags) Other issues: – restrict advertising, especially if directed at children or for certain status goods – restrict the establishment of malls and the opening hours of shops
Planning	Mobility: – improve conditions for pedestrians, cyclists and users of public transport in cities (dedicated lanes, dedicated programming of traffic lights, etc.) – make the use of cars in cities less attractive (e.g. by reducing parking space) and more expensive (e.g. with congestion charges)

242 Note that the table just contains an inventory of the ideas under discussion in the literature cited above (see beginning of section 5.1.2) and is not a proposal that I intend to advance here.

243 "Obsolescence" refers to the early failure of products long before their usual technical end-of-life is expected.

Category	Sufficiency policy instruments
	– create parking space specifically for car sharing schemes – support the "compact city" of short distances, mix living and working areas Food: – support urban farming to decrease land use and transport needs Housing/buildings: – reduce new construction in the countryside – share public buildings (e.g. schools) and spaces among different uses in order to increase the intensity of use – help elderly people to move into smaller homes if they want Appliances/products: – (no specific instruments proposed in this category) Other issues: – provide "consumption-free" non-commercial areas such as benches in pedestrian areas
Economic instruments	Across all consumption areas: – carry through (more resolute) ecological tax reform to increase prices of energy and resources and decrease labour costs to foster repairing and resource-extensive services – tax carbon dioxide emissions – introduce impact caps and cap-and-trade schemes on various levels (e.g. personal carbon trading for individuals or communal tradeable land certificates in order to reduce land use) – introduce subsidies for collective forms of use Mobility: – higher taxation of fossil fuel energy, in particular taxation of air traffic and ship fuel oil – introduce a fee on air transport – remove environmentally harmful subsidies leading to increased transportation (e.g. commuter subsidies) – introduce subsidies for environmentally sound alternatives (public transportation)

Category	Sufficiency policy instruments
	Food: – tax meat and dairy products (except if from organic farming) – tax fast food – remove environmentally harmful subsidies (e.g. subsidies for factory farming) Housing/buildings: – energy prices: put a higher burden on relatively high consumption (progressive tariffs or taxes) Appliances/products: – tax selected environmentally unsound products (e.g. batteries, plastic bags, or certain luxury goods) – financial incentives for discarding old appliances (or cars) under the condition that no new ones are bought – reconsider depreciation rules – remove subsidies leading to increased consumption Other issues: – tax advertising – tax land use (or soil sealing)
Nudging	Across all consumption areas: – make green choices more easily accessible or salient – social comparison: provide information on green behaviour of others Mobility: – (no specific instruments proposed in this category) Food: – voluntary vegetarian days in canteens – offer vegetarian dishes as default with the "opt-out" possibility of ordering meat – reduce plate size in restaurants or buffets to avoid food waste Housing/buildings: – mandatory installation of water meters or smart energy meters – installation of a crystal ball that glows red when energy use is high – present own energy or water use compared to that of others on utility bill – apply "click" stickers to light switches

Category	Sufficiency policy instruments
	Appliances/products: – place environmentally unsound products in unfavourable positions on the shelves in a store Other issues: – mandatory default settings (e.g. emails instead of paper mailing, two-sided printing)
Coopera-tion	Across all consumption areas: – support existing (e.g. libraries) and the establishment of new institutions which promote practices of sharing, renting, reusing, repairing etc. ("social innovations", cf. section 3.2.5) – stimulate public consumption at the expense of private consumption in order to create more public goods (no specific instruments proposed in this category for the different consumption areas)
Information	Across all consumption areas: – communicate "best practice" sufficiency examples to specific target groups – increase awareness of the standards of luxury that we have attained – increase awareness of the legacy of values and the mechanisms that stimulate consumption Mobility: – (no specific instruments proposed in this category) Food: – inform consumers about advantages for health of reducing meat consumption Housing/buildings: – (no specific instruments proposed in this category) Appliances/products: – shift from (energy) efficiency labels for appliances to primarily communicating absolute energy use Other issues: – (no specific instruments proposed in this category)

5.1.3 Discussion

Having ordered and presented the collection of policy instruments (Table 9), each policy category will be discussed briefly in the following. As it is not possible here to analyse each of the proposed policies in detail, a more in-depth analysis will be provided in the next section for selected policy examples (section 5.2). The aim here is just to roughly point out the strengths and weaknesses of each category of policy instruments and prepare the ground for the selection of the examples. The discussion addresses the aforementioned criteria for the evaluation of environmental policy instruments (effectiveness, efficiency, likelihood of implementation deficits, political feasibility, and impact on individual liberty) as far as these are applicable, and highlights further aspects that are salient in each category.

Regulatory law

Regulation ("command and control") is the most widespread and traditional means of environmental policy and is responsible for a number of success stories in environmental protection (Jänicke 1996: 12, 2003: 102, Jänicke/Jörgens 2009: 156). The regulatory approach is indispensable where direct environmental harm needs to be countered (Jänicke 2003: 102). However, it also has a number of weaknesses: not only is it seen as economically inefficient, but it often also has only limited acceptance and meets the resistance of those concerned by a regulation, with the result that the implementation can be delayed or deficient (Jänicke 1996: 12, 2003: 103). For example, from the policies in Table 9, design requirements on product lifetimes are feasible and useful in some cases, but can sometimes create rather high costs for manufacturers and require quite intensive administrative surveillance procedures[244], and thus they have rather low economic efficiency. Other proposed instruments would, however, not substantially increase the existing administrative burden as these things are regulated and enforced already and the standards would just be raised (e.g. for speed limits, carbon dioxide emissions from cars, limits on product energy use). Implementation deficits are possible across all the instruments proposed, and are likely in particular for stricter standards on food production, not only because of the sheer number of producers, but also because of the difficulties of controlling imports. In terms of effec-

244 Such requirements and according problems are discussed, for instance, in the context of the implementation of the European Ecodesign Directive (cf. Spengler 2016 b).

tiveness, there are some instruments that may, depending on their concrete design, make very significant contributions to sustainability, for instance the already mentioned upper limits on carbon dioxide emissions of cars and environmental standards for food production. Other proposals, however, are suspected of being rather "symbolic policies" (cf. Newig 2007), such as bans on outdoor heaters or restrictions on advertising for status goods.

A crucial issue for regulatory policies in the consumption context is their political feasibility and public acceptance. If the aim is to reduce consumption levels directly, setting regulatory standards would likely meet strong resistance from consumers (OECD 2002 b: 19), as the environmental impacts of final consumption are mostly indirect, while the approach of regulation is, as mentioned above, commonly rather used to prevent direct environmental harm. There are indeed examples where consumption levels have been directly regulated in times of scarcity, for example those of water or energy, for instance through bans on watering the garden or restrictions on car use (OECD 2002 b: 19). However, where such directly perceivable scarcities do not exist, direct regulation of consumption levels would put too strict limits on individual liberty, and the commensurability principle would most likely require selecting a different, less coercive means (cf. section 4.3.5). As Table 9 shows, direct restrictions are indeed among the proposed sufficiency policies (e.g. prohibit driving private cars on Sundays), but most instruments assigned to the category of regulatory law do not directly regulate consumption levels. Instead, they seek to establish standards on consumption goods that do not inhibit obtaining or using these goods, but rather have an impact on their price or certain properties. Thus, for instance, setting an absolute limit on the carbon dioxide emissions of cars would result in some bigger cars not being available for private use anymore, and environmental standards on food production would result in higher food prices. These policies would certainly still have to struggle for public acceptance, but probably not as much as direct restrictions on car use or, say, meat consumption.

The instrument from this policy category which would be most interesting to analyse in detail in the light of sufficiency policy is, in my view, the idea of introducing stricter environmental standards on food production, as it would have significant effects and touches, at least in the political discussion in Germany, upon highly topical issues. However, political measures in the sense of sufficiency policy with regard to food consumption, including measures that restrict unsustainable choices by way of regulatory law, have already been discussed comprehensively elsewhere (cf. Voget-Kleschin 2012: 231 ff.). I therefore do not include an example from the category of regulatory law in the detailed discussion in section 5.2.

Planning

Instruments in this category include overarching planning instruments such as regional development and land use plans as well as specific sector plans, for instance on air quality control, waste management, and water management (Kunig 2003: 208 f.). Planning determines what will be possible or allowed, for instance, in a certain area, thus having a clear influence on human activities within this area. Many planning processes lead to the creation of infrastructures that will be in use for a long time, so that changes are not possible or too expensive in the short term. In particular, once an area of land has been assigned for a specific purpose (industry, living area, park), this determination may be set for centuries.

Environmental instruments from the category of planning are usually not discussed as alternatives to other instrument categories[245], but rather taken as a necessity. Planning instruments are thus rarely assessed by means of the criteria used here (efficiency, effectiveness etc.) in the same way as instruments from the other categories. Approaches to planning have also not been taken up systematically by the literature on sufficiency so far (e.g. with regard to air quality, waste and water management). As Table 9 shows, only the use of land and in particular the creation of mobility infrastructure are present with a number of instrument proposals in the literature on this subject (cf. also Schneidewind/Zahrnt 2014: 75 f.). From the instruments in Table 9 that are not primarily related to mobility, the one that would have the most significant effect is probably the idea of reducing new construction in the countryside. The idea, however, is at odds with typical communal development aims and would thus very likely encounter problems of political feasibility. The listed instruments related to mobility are all well known, and many cities have already adopted one or another of these measures by now. Often bundles of measures to make mobility patterns more sustainable are taken. However, there are few

245 Unlike the other instrument categories, which are frequently compared to each other in the literature (regulation vs. economic instruments, regulation vs. cooperation, etc.).

places where this has been implemented consequently in the sense of sufficiency policy. As mobility is one of the major discussion fields in the context of the sufficiency debate, a more comprehensive analysis of planning instruments in relation to transport infrastructure will be carried out in section 5.2.3.

Economic instruments

Economic instruments can be divided into three categories, as indicated in Table 8—taxation and fees (public revenue), subsidies and tax reliefs (public spending), and cap-and-trade mechanisms (public revenue neutral). Economic instruments are generally said to be a particularly efficient way of dealing with a problem, as economic actors flexibly adapt to changed conditions (OECD 2002 b: 17 f.). Reductions of environmental impacts are not only said to be allocated efficiently, but, by means of adjusting the level of a fee or tax or by setting a cap, will also result effectively in the desired level of reduction. However, as regards the consumption context, a limitation to the effectiveness of economic instruments is that prices are not the only factor that influence consumers' choices (OECD 2002 b: 18). Thus, increasing prices of certain luxury goods, for instance, will probably have a limited influence on someone's decision to purchase such goods.

Instruments that redistribute money also have potential negative side effects, depending on the degree of redistribution. This includes implementation deficits, even though they are associated with economic instruments to a lesser degree than with regulatory law. Measures that involve public spending (subsidies) are rather easy to enforce; however, their effectiveness is limited because they often lead to mainly windfall profits and because those causing external costs are not charged (Jänicke 2003: 104). An example from Table 9 of such a distributive instrument are financial incentives for discarding old energy-using products. However, in the context of sufficiency policy, it is rather the removal of subsidies that is advocated in order to decrease incentives to consume.

A negative side effect in particular of revenue-raising instruments is the potentially relatively high financial burden on poor households (OECD 2011 a: 141). While economic instruments normally only interfere with individual liberty to a limited degree, as freedom of choice is still given, a high financial burden could turn into real restrictions on liberty for some social groups. It is however possible to attenuate such effects through redistribution measures. Potential negative social and economic impacts of revenue-raising instruments will determine their political feasibility and public acceptance, and are also a central concern for revenue neutral instruments such as cap-and-trade schemes, since these also directly redistribute money.

Most of the economic instruments proposed (cf. Table 9) are well known and there is already considerable practical experience, but in order to qualify as sufficiency policy more resolute implementation of these policies would be necessary. Taxes, for instance, are proposed in the literature as an instrument of sufficiency policy across all consumption fields and thus appear to be an important measure in this context. They are proposed for environmentally significant goods, such as meat, fast food, plastic bags, and batteries, as well as more generally for resources such as fossil fuel energy and emissions such as carbon dioxide. Environmental taxes will thus be discussed in detail in section 5.2. Another economic instrument that is often mentioned in the context of sufficiency, but which has not been enacted as a policy so far, is a cap-and-trade mechanism, namely personal carbon trading. As the approach of this instrument is particularly interesting for sufficiency, it will also be dealt with in detail in section 5.2.

Nudging

The debate on "nudging" is relatively new and the concept has only gained attention within the environmental discussion in recent years. A nudge is "any aspect of the choice architecture that alters people's behavior in a predictable way without forbidding any options or significantly changing their economic incentives" (Thaler/Sunstein 2008: 6). Regulatory and economic instruments should thus not be considered as nudges, even if they influence people's choices, too. However, as opposed to the exclusion in this widely accepted definition, economic incentives are often mentioned as examples of nudges[246]. Further, instruments that are labelled as nudges often consist in the provision of information, which is a tradi-

246 See, for instance, Smeddinck (2011: 389 ff.) and cf. the discussion in Ölander/Thøgersen (2014: 353). Note that even Thaler and Sunstein mention examples of nudges in their influential book from which the definition is taken that actually consist in changing economic incentives. In a chapter on environmental issues,

tional means of environmental policy (see below). For instance, the very successful European Energy Label is an information instrument, but as it also aims to shape the choice architecture towards more environmentally sound purchase decisions, it is widely considered a nudge, too (cf. Ölander/Thøgersen 2014: 345, Reisch/Sandrini 2015: 97). There is thus a considerable overlap with this instrument category, and Table 9 only lists instruments as nudges that are not primarily information instruments. What is new with regard to nudges is the focus on the results of behavioural research and the according recommendations for adaptation of the choice situation to increase the share of people that will behave as intended. Despite these overlaps, there are several examples of original nudges in the environmental realm that do not fall into either of the other policy categories, for instance the purposeful setting of default options, placement of products in shops, or confronting people with targeted data about the behaviour of others (social comparison).

Nudges are said to be particularly suitable in the context of consumption policy, especially in cases where the more traditional instruments do not work properly or would obtain limited acceptance because of too strong interference with individual liberty or other negative side effects (Reisch/Sandrini 2015: 22). A major advantage of nudges, according to their advocates, is that freedom of choice is still given, since unlike regulatory approaches they do not alter the set of available options. They are thus characterised by a low degree of interference and costs are also low, so their political feasibility is, in principle, rather high. Implementation deficits are therefore not an issue in the literature on nudges. In addition, some nudges are, according to empirical studies, highly effective, even though it is difficult to anticipate their effectiveness (Croson/Treich 2014, Sunstein/Reisch 2014). Disadvantages of nudges consist, above all, in worries about the paternalistic character of influencing individual choices silently. However, as pointed out in section 4.2.1, there is no instance of paternalism if a nudge aims at reducing potential harm to others (as is usually the case in environmental affairs), and not only to oneself. Of course, it can still be seen as problematic if consumer choices are influenced secretly for a specific purpose, instead of openly discussing and informing consumers on the reasons and according policies taken. On the other hand,

---

they sketch ways to influence people's choices by means of nudges. According to them, the most important way is to set economic incentives through environmental taxes or cap and trade systems (Thaler/Sunstein 2008: 183 ff.).

consumer choices are influenced silently from many sides, including for the purpose of increasing consumption.

There are a number of nudges that could be used to decrease consumption levels (Table 9), especially in the consumption areas of food and housing, and some that apply more generally. The examples related to food all apply to the specific situation of eating in a canteen or (similar type of) restaurant and do not work with regard to shopping for food or meals eaten at home. The examples related to housing, directed at reducing energy and water use, are in part already practised widely. However, to qualify as real nudges, they could often be designed better to influence people's behaviour more effectively. For instance, the comparison of one's own energy use with that of one's direct neighbours may, according to a study, reduce energy use by 2%. However, such social comparison can also have the opposite effect if someone's behaviour is already "greener" than average (Reisch/Sandrini 2015: 105). Further, default settings and making green choices more salient (and decreasing the visibility of unsound choices) are also possible instruments to nudge people into behaviour that is more in line with sufficiency.

However, overall the scope of application of original nudges—that is, excluding economic incentives and plain information instruments—that are appropriate in order to steer people towards sufficiency seems to be rather limited. There are, for instance, currently no concrete ideas on how nudges could be applied to choices regarding mobility[247]. With regard to this area, but also in other decision situations, the most effective nudge default setting is often not possible. When people choose from a range of options offered by a market, as is often the case with regard to choice situations that affect the environment, no default rule is available but only other nudges that make green choices more salient (Sunstein/Reisch 2014: 150 ff.). In particular with regard to the option of *not* buying something or consuming less, it is difficult to think of ways to set this as default, apart from the often-cited example of double-sided printing. Also, more generally with regard to nudges, it seems unclear how, for instance, the sheer

247 Sunstein and Reisch (2014: 152) mention attractive bicycle lanes and well-maintained public transportation systems in the context of choice architecture. However, I think that such measures, which I have listed under the category of planning, consist in the improvement of certain options rather than in changes to their presentation and thus do not match the definition of nudges, or at least involve more than the mere nudge.

amount of physical goods someone buys or the distance someone travels per month should be influenced this way. Even if there are a number of successful approaches, the effectiveness of nudges for the purposes of sufficiency appears to be limited, and thus no example from this category will be discussed in detail in section 5.2.

Cooperation

Cooperation as a policy category as conceived by Jänicke comprises different sorts of negotiations and agreements between those causing environmental pollution and the state or environmental organisations (cf. Table 8; Jänicke 2003: 105 f.). The role of the state is reduced to signalling the possibility of establishing voluntary commitments to polluters or entering into negotiations as an alternative to direct regulation, and it is the task of the polluters to elaborate measures to deal with an environmental problem. Much literature is dedicated to the evaluation of cooperative solutions, such as voluntary agreements in the environmental field (cf. OECD 2003, Bracke et al. 2006, Jänicke/Jörgens 2009: 172 ff.). Cooperative instruments are supposed to have advantages over direct regulation as they are more likely to be accepted by the addressees, can react more flexibly to changes to the status quo, and entail only minor restrictions to the freedom of the addressess as they leave open the detailed approach to achieving a target. On the other hand, they are often criticised as rather ineffective and inefficient, since the targets agreed with industry are not always ambitious, monitoring mechanisms are often insufficient, and negotiations are often time-consuming (OECD 2003: 14, Spengler/Dick 2007, Jänicke/Jörgens 2009: 173). Still, they can be useful as supplementary mechanisms to other instruments (Jänicke/Jörgens 2009: 174).

In the context of this policy approach, the polluters are companies, that is, producers. Since sufficiency policy first of all targets consumption, the approach of cooperation does not seem to have a place in the discussion on sufficiency policy. However, it is possible to transfer the idea of cooperation to the consumption sphere and interpret sharing schemes and other "social innovations" (cf. section 3.2.5) as cooperation instruments. Cooperation here takes place between consumers who voluntarily organise the sharing, exchange, or repairing of goods in order to reduce the environmental impact of their consumption habits. Similar to such social innovations, the idea of providing more public goods at the expense of private goods can be seen in the light of cooperation (cf. Table 9).

Still, there are some differences between the kind of cooperation between producers and the state as originally referred to by Jänicke's category and the "social innovation" kind of cooperation between consumers, which is why the assessment of cooperation with producers in terms of efficiency, effectiveness etc. cannot be directly transferred to consumer cooperation. These differences include no negotiations taking place, as the aim is not to cover a large part of the "market"; there is no competition between consumers (at least not in the same sense as for producers); consumers are numerous and not organised like corporate actors; and the state does not necessarily play a role at all, since consumers organise themselves. What can nevertheless be said in terms of effectiveness is that even if supporting such schemes from the side of the state is certainly helpful in mainstreaming these practices, cooperation again relies on the voluntary uptake of sufficiency practices. The effectiveness of this approach is therefore limited (cf. section 4.1.1) and, like the kind of cooperation with producers, it should be seen rather as an ancillary approach in addition to other instruments.

Information

Providing environmental information to consumers through product labels, information campaigns, individual energy audits, or similar measures is important in order to enable them to integrate such information into their decisions. Introducing environmental information into education programmes (cf. e.g. OECD 2008: 25), such as school curricula, also belongs to this instrument category. Informing consumers is characterised by a low degree of interference, as, like nudging, it does not decrease the number of available options or change their price incentives. Public acceptance therefore usually does not pose significant problems to an information instrument, apart from certain drastic examples[248]. The most widely discussed issue with regard to information as an environmental policy instrument is its effectiveness. Some information instruments, for instance the European

248 Consider, for instance, the requirement to display photographs of certain diseases and according messages on cigarette packages. Such approaches certainly also qualify as nudges (see above).

energy label for products, have proven very successful in steering purchase decisions towards more environmentally sound products (Molenbroek et al. 2014). Information is also important as a complementary approach to other instruments, such as taxes, as it can help to achieve a higher degree of acceptance for such less "soft" policies (OECD 2011 a: 141). However, according to evaluations, the effectiveness of information provision alone, in particular through campaigns, is often limited as the increase in knowledge often does not result in according behavioural changes (OECD 2002 b: 20, Rubik et al. 2009: 41, Ölander/Thøgersen 2014: 342 f.), although there are differences in effectiveness depending on the type of information provided and the context into which it is embedded (Delmas et al. 2013).

Most of the ideas on advancing sufficiency mentioned in the category "information" in Table 9 would have the form of the rather weak instrument of information campaigns (e.g. informing consumers about advantages of reducing meat consumption for health, increasing awareness of luxury and consumption drivers). The proposal of communicating "best practice" sufficiency examples and experiments to specific target groups is slightly different: this information instrument does not directly appeal to the individuals addressed to change their behaviour, but it informs consumers about what alternative forms of action are possible and are successfully practised by other individuals. This might encourage people to adopt different consumption patterns, too. Again a different approach, namely that of product labelling, is used by the idea of shifting from primarily communicating the energy efficiency of a device to highlighting its absolute energy use. Given the success of the energy label for products, this might entail a more significant effect than a campaign.

Overall, information is the instrument characterised by the lowest degree of interference. Like the aforementioned instruments of cooperation, the effectiveness of this instrument relies again on individual agency, as it depends on whether individuals take the information provided into account in their decision-making or not. The framework conditions of consumption decisions are only altered in a limited way by the approach of information (e.g. in the case of product labels), or even not at all in the case of campaigns (Rubik et al. 2009: 41). Therefore, information should rather be seen as a complementary approach to price incentives and improvement of the "hardware" of products, services and infrastructure (OECD 2002 b: 21).

5.2 Selected policy examples

Three examples of policy instruments that can support consumption patterns in accordance with sufficiency—environmental taxes, personal carbon trading, and planning instruments for urban infrastructure to reduce motorised individual transport—are presented and analysed in the following. These policy examples all meet the requirement that they are based on the harm principle, and thus rely on a justification that is neutral with regard to different conceptions of the good (cf. sections 2.1.3, 4.4.1). They have been selected because they represent the three categories of solutions discussed at the end of chapter 4 (cf. section 4.3.6): targeting certain resources or pollutants, focusing directly on individuals, and targeting certain consumption practices that are problematic from an environmental point of view. Further, they are interesting as their potential effects are rather broad and significant. Finally, they clearly match my definition of sufficiency policy (cf. section 5.1) and are frequently mentioned instruments of environmental policy, so a discussion of them in the light of the subject of this study adds an interesting perspective.

The following analysis includes for each policy a description of how this policy works, its (expected) effects in the sense of sufficiency, as well as its general advantages and disadvantages regarding the five criteria used before (cf. sections 5.1.1, 5.1.3)—effectiveness, efficiency, likelihood of implementation deficits, political feasibility, and the impact on individual liberty—which, as explained above (cf. section 5.1.1), at the same time represent the theoretical background developed in chapter 2.

5.2.1 Environmental taxes on resources and emissions

As mentioned above, environmental taxes are a well-known instrument of environmental policy and are already implemented in many countries. Resource prices usually do not include the external costs emanating from the extraction of a resource that fall on society. Instead, sometimes resources such as water and energy have been or are still subsidised by national governments (Weizsäcker et al. 2009: 316). Emissions have not had a price at all for most of the time. Environmental taxes are therefore levied on goods or resources that cause adverse environmental impacts in their production, use, or disposal, or directly on harmful emissions. Most often, such taxes target harmful emissions, fossil energy carriers, electricity, water, or waste, but they also exist for products such as plastic bags, batteries and disposable cameras (cf. Ekins 1999, Bosquet 2000, OECD 2008: 14; cf. also Table 9). According to economic theory, for optimum efficiency an environmental tax should be directly linked to the substance or resource use that is causing environmental harm (cf. section 4.3.6). However, sometimes this is not possible or desirable, so products or other inputs to production are used as proxies (Ekins 1999: 47). The essence of environmental taxes, as David Pearce has pointed out, "is to encourage tax avoiding behaviour" (1990: 375). This usually means that economic actors decrease their demand for the substance or resource taxed, given free and properly working markets. Sometimes, as will be discussed below, it can also mean that they find other ways of "avoiding" a tax. A second aim is to generate tax revenue for national households; sometimes, environmental taxes are coupled with reliefs on non-wage labour costs in order to make them revenue neutral. In this case, they do not only have the steering effect of decreasing the demand for certain resources or products, but they also, ceteris paribus, increase the demand for labour. The revenue raised with the tax can, instead of lowering other costs, also be used for environmental aims[249].

Weizsäcker et al. (2009: 313 ff.; cf. also Weizsäcker et al. 1995: 219 ff.) propose a long-term environmental tax in order to raise the prices of certain resources steadily in proportion with resource productivity increases[250]. This way, production costs would remain constant, while resource inputs would decrease over time. They suggest that the system should be "made binding for 50 years or more, with fairly tough clauses for exemptions or deviations from the rule" (2009: 327). It therefore provides planning reliability to economic actors and unlike taxes that are introduced at once does not have the problem that it is either high from the beginning and will be a significant burden for taxpayers, or too low to have relevant effects (Weizsäcker et al. 2009: 320).

An environmental tax on certain resources that increases steadily has the important advantage that it prevents rebound effects from resource

249 See the classification of environmental taxes in Ekins (1999: 42 f.).

250 Weizsäcker et al. actually seem to suggest that the tax should also be used to smooth out price fluctuations caused by economic developments and financial markets, referring to the example of China, where "energy prices have been fixed by the state for decades", "without much debate or a need to justify [this] decision against the fiction of a 'free market'" (Weizsäcker et al. 2009: 319).

productivity increases, so that not only environmental efficiency but also sufficiency is supported by this instrument (Weizsäcker et al: 2009: 328). Most often, environmental taxes primarily concern production facilities, but there are also taxes, for example, on energy, that concern both the production side and final consumers. Further, if a tax is levied on production facilities (e.g. on waste water discharges) the increased costs of production will often be passed through to consumers, so there will also be an indirect effect on consumption. Environmental taxes thus make commodities more expensive, give incentives to reduce their consumption and are thus expected to lead to a reduction in the according environmental impacts. However, apart from these changed economic incentives, environmental taxes do not per se alter framework conditions that are supportive of individual sufficiency practices. For instance, they do not provide additional information on environmental impacts of consumption goods or change the available infrastructure (cf. section 5.2.3). Nevertheless, the revenue from environmental taxes can be used in addition to support individual sufficiency practices economically.

The effectiveness of environmental taxes in terms of reducing environmental impacts depends on the object of the tax. For demand price-inelastic resources the effects are limited, but if the revenue from the tax is used in addition to promote environmental aims, the effects are greater (Bailey 2002). For carbon emissions in particular, significant reductions can be achieved with environmental taxes (Bosquet 2000). Also, environmental taxes seem to be more effective in reducing pollutant emissions than reducing energy use (Morley 2012), which is in accordance with the findings about price elasticity, as energy is a rather price-inelastic commodity due to its importance for industrial production and our daily routines. This points to a general disadvantage of environmental taxes: even if they negatively influence demand by means of additional costs, they do not imply impact caps and thus leave it more or less open whether a reduction target is reached or not (Weizsäcker et al. 2009: 315). Of course, an approach based on taxes could be designed in such a way that a certain target is approximated, and the tax rate could be adjusted after some time. Still, other instruments such as cap and trade systems (cf. next section) are generally more appropriate in reaching a clearly defined target. A further limitation to the effectiveness of environmental taxes is the typical lack of international cooperation regarding taxes. If an environmental tax is imposed nationally, production may relocate to other countries or national production will not withstand international competition, and the effects of the tax will be reduced (Ekins 1999: 56 ff.). The extent to which this happens depends on the mobility of capital (Bosquet 2000: 28) and the level of the tax. In the case of taxes that directly target final consumers, for example taxes on electricity or water, this problem can usually be disregarded, as consumers are unlikely to relocate elsewhere because of a (moderate) tax. However, in some cases, where it is possible to acquire a resource beyond the borders of a state, this point may be relevant, as in the case of "fuel tourism".

As regards efficiency, economic instruments of environmental policy, such as environmental taxes, are generally considered to be the best solution. Producers can decide whether to continue polluting (or using the resource taxed) and pay the tax, or to invest in pollution abatement (or more resource efficient) technologies, or even stop producing. Those with the lowest costs of decreasing their emissions of the pollutant taxed (or use of the resource taxed) will choose to emit less, while others will not unless the tax becomes higher. According to economic theory, this leads to environmental improvements at overall minimal cost (Ekins 1999: 42, Felder/Schleiniger 2002: 108). In principle, the same also applies to consumption, except that the costs include not only monetary, but also emotional "costs" due to altered choices. The efficiency of the instrument, however, is in practice reduced by the widespread use of exemptions and lower tax rates for certain economic sectors because of fears about international competitiveness (Ekins 1999: 57, Ekins/Speck 2000: 108, Felder/Schleiniger 2002). Such exemptions are a common concession to political feasibility, even if there appear to be better ways to deal with the problem of competitiveness, for instance through differentiated tax revenue redistribution (Felder/Schleiniger 2002).

Besides worries about competitiveness, the political feasibility and public acceptance of an environmental tax depend on its distributional effects (Ekins 1999: 58 ff.). As basic commodities such as energy or water tend to be relatively more important in the expenditure of low income households, a tax on these could create social hardship. However, various compensation schemes are possible, including the aforementioned option of making environmental tax revenue neutral by lowering non-wage labour costs. Other options are to introduce a tax-free threshold or make the tax progressive, even if this may again be complicated in implementation. According to a survey, a properly designed carbon tax system could be acceptable to 70% of the public (Bristow et al. 2010).

In order to be effective in steering consumer behaviour, taxes need to be set at a sufficiently high level. However, beyond a certain point, a tax may suffer from implementation deficits: those affected by it may try to avoid the tax, if possible, by acquiring the resource or good taxed via newly emerging black markets, or, as mentioned above, by acquiring it beyond national borders where it is not taxed in case of a lack of international harmonisation. Further, both producers and final consumers might start to look for alternatives and substitute a resource with a similar one that is not being taxed. For instance, if oil and gas were taxed significantly, some people may start cutting down trees in the next forest to get firewood for free. The overuse of the replacement resource and potential environmental effects of such replacement should be anticipated, as the aim of the tax is not to shift environmental impacts but to decrease them (FoEE 1995: 180). This latter point does not appear to have been a major issue so far, as it is reflected little in the academic literature. However, as said before, if tax rates were higher in order to have a greater effect in terms of suffiency, this might become relevant.

As regards their consequences for the liberty of individuals, environmental taxes are in principle a solution with limited impact. By internalising external costs they impose the costs of environmental harms on the agent who profits from them, including the possibility that the agent abstains from an activity if the costs are too high. Like "nudges" (see section 5.1.3) and unlike solutions that focus on specific consumption activities (see sections 4.3.6, 5.2.3), apart from the related costs, they do not forbid any of the available options, so freedom of choice is still given, provided that someone is able to pay the higher costs. This qualification, however, points to the distributional effects of taxes, which can have adverse effects in terms of liberty. As mentioned above, it might be necessary to set an environmental tax at a rather high level, which will negatively affect poorer groups of society and decrease their freedom disproportionately unless effective compensation measures are taken. A very high tax could even lead to a situation that is not very far from a ban, but it is unlikely that resources or pollutants would be taxed in such a way. Also, the mechanisms needed to enforce a tax pose a possible threat to liberty. This depends on their extent of course, but some surveillance to make sure that a tax is really paid for the goods on which it is levied is always needed. In addition, in practice it is impossible to internalise all external environmental costs by means of taxes, so necessarily a selection has to be made among the most relevant resources and pollutants. Therefore, some unfairness with regard to the tax burden is always involved, as it may be rightly asked why a certain resource is taxed if another one, which is also environmentally relevant, is not.

5.2.2 Personal carbon trading

With regard to carbon dioxide emissions, one of the currently most pressing environmental issues (cf. chapter 1), another economic instrument besides environmental taxes, namely the idea of individual pollution allowances has been subject to comprehensive research. Different names and variants have been proposed in the literature on the subject (cf. Seyfang et al. 2009, Fawcett 2010, Starkey 2012 a, 2012 b). A common umbrella term for these is "personal carbon trading", which I will use in the following.

Similar to existing emission trading systems for industry—that is, "upstream" in the production chain[251]—, personal carbon trading establishes a system of allocating emission rights to individuals and a platform that allows them to trade these. Basic elements to personal carbon trading are that national (or international) emissions are subject to a cap which decreases over time; the total emissions or a certain share of the total (depending on the particular scheme, after subtracting a part reserved for industry and government authorities) is distributed via some mechanism to individuals as emission rights; these emission rights need to be surrendered, for instance, whenever an individual pays a private energy bill, buys a flight ticket, or fills a car with petrol; the emission rights are tradable (Seyfang et al. 2009, Fawcett 2010). Individuals who have an energy intensive lifestyle so that the emission allowances allocated to them are not enough would have to buy additional emission rights at the market price from others who have surplus allowances. The initial allocation usually proposed consists in equal shares for all individuals. As trading the initially obtained emission rights allows for a diversity of lifestyles (at least for those who are able to pay the price of additional allowances) and emissions can be saved where the saving costs are comparably low, the instrument is seen not only as a fair, but also as an economically efficient way to reduce carbon dioxide emissions.

251 For a comparison of upstream and downstream emission trading systems, see Defra (2008: 56 ff.).

A significant difference between the various proposed schemes is to whom emission allowances are allocated: only individuals, or government agencies and industry as well. For simplicity, I understand personal carbon trading as referring to a scheme comprising emission allowances for individuals only, as they are the primary addressees of this instrument. A further difference relates to the types of carbon dioxide emissions that are covered. While some schemes only encompass the emissions from household energy use, others include those from transport as well. In theory, personal carbon trading could even encompass the majority of all carbon dioxide emissions, that is, all emissions that are caused somewhere in the chain leading to private consumption in the end. However, establishing a system that mandatorily requires the information for all, including imported, products about their embedded carbon dioxide emissions would be extremely costly and the system would be prone to fraud, as monitoring the correctness of the information would be very difficult (and add further costs). Thus, as a concession to feasibility, personal carbon trading generally does not encompass the indirect emissions from production that are linked to individual demand (Seyfang et al. 2009: 4). I will understand personal carbon trading here as encompassing the greatest possible share of individual carbon dioxide emissions given this feasibility constraint. For the UK, researchers estimate that including household energy use and personal transport emissions, personal carbon trading would cover 42% of all carbon emissions from energy use (Fawcett 2010). As only a part of carbon dioxide emissions would be covered, personal carbon trading researchers emphasise that it would not be a standalone policy (Fawcett 2010: 6870). Thus, despite the (overly optimistic) expectations of some (cf. e.g. Müller 2011), the instrument would have to be integrated into the existing policy mix and replace only some of its components.

To realise sufficiency, personal carbon trading is in principle the perfect policy instrument: it directly implements sufficiency in the form of a cap on average individual emissions (thus avoiding "excessive" consumption, cf. section 4.3.3) and thus requires individuals to adapt their consumption pattern so as to stay within the limit or to buy additional pollution allowances; in addition, it allows individuals to monitor their own emissions and learn about the carbon intensity of different types of behaviour. Besides effects in the sense of sufficiency (demand reduction), personal carbon trading also supports the adoption of technical solutions, such as the replacement of an electric device with a more energy efficient one, to decrease personal carbon emissions. Given this visibility and feedback of carbon emissions and the knowledge about having a limited allowance which will further decrease, the policy activates not only an economic motivation but may also have further psychological and social effects (Fawcett et al. 2007, Fawcett 2010). However, given the practical limitations in terms of the scope of carbon emissions that can be covered (no embedded emissions), the cap is not complete. Further, the instrument is unlikely to be feasible for many other environmental impacts (except e.g. for water), given that most of these are, at the level of final consumption, embedded in goods and services rather than directly caused by individuals.

One of the crucial advantages of cap and trade mechanisms is their supposed effectiveness since they start by fixing a pollution (or resource use) target which by definition is not surpassed. Taxes, by contrast, can only indirectly work towards reaching a quantitative target by modelling the tax rate in a way that demand can be expected to decrease until the predefined target. Thus, in theory, if all sectors of the economy were covered, there would be no rebound effects under this instrument, apart from international effects if applied only nationally. In practice, however, adjusting the certificate supply to a quantity that will effectively lead to a decreasing demand is not a given, as there may be an unexpected oversupply of certificates, as experience with other systems, such as the European Union industry greenhouse gas emissions trading system, shows (Seyfang et al. 2009: 12). The price system of emission allowances may also be distorted by speculation, which might both hamper its effectiveness and entail problems for investors and poor households (Weizsäcker et al. 2009: 315). Further, the effectiveness of the policy can be compromised as the scope of personal carbon trading is limited to individuals and emissions from energy and transport, as a shift towards other economic sectors, which are not included, is possible. This might lead to a microeconomic "carbon leakage" effect, for instance by people taking activities that need electricity to the office in order to reduce their use of it at home. Potential implementation deficits (see below) further add to the aspects that may decrease the effectiveness of the instrument.

An advantage of cap and trade mechanisms in general, according to economic theory, is their high economic efficiency. In a similar way to environmental taxes, the trading mechanism allows for realising the abatement of environmental impacts where costs are low. However, those who are enthusiastic about the efficient allocation of emission rights by the market often forget that the mechanism itself comes at a cost[252]. According to estimates, the costs of running a system of personal carbon trading would be very high and exceed the costs of other instruments, such as an upstream cap and trade mechanism or taxation (Defra 2008, Fawcett 2010: 6871, Starkey 2012 b). According to the available studies, the high economic efficiency of the abatement mechanism itself would therefore be more than cancelled out by its running costs.

The potential implementation deficits are similar to those of environmental taxes: black markets, people acquiring energy resources beyond national borders if possible, and shifts towards commodities that are not included in carbon trading may be the outcome (see section 5.2.1). Consumption researcher Maurie Cohen (2011) has in particular described worries about the potential emergence of black markets if a policy of personal carbon trading should be implemented: Comparing it with rationing policies in times of war, he states that there is sufficient evidence to conclude that putting external limits on consumer demand "inevitably leads to unlawful activity and black markets spring up whenever and wherever auspicious opportunities exist" (Cohen 2011: 213). As Cohen describes, various consumption goods (mainly food, clothing, and fuel) were rationed in the UK during World War II and could be obtained only in exchange for coupons. This resulted in a significant degree of violations, including forging and stealing coupons, pretending to have lost one's ration book and applying for a new one, transferring goods obtained for commercial purposes to private use, under-the-counter sales, and illicit production of food items (e.g. illegal slaughter). Coping with such "innovative commerce" would require significant policing action, which is still unlikely to completely eradicate illegal activities (Cohen 2011: 217). Cohen's comparison, however, does not take into account with regard to the illegal activities in the past the perspective of, first, current technical options that did not exist in the 1940 s (i.e. a digital transaction system) and, second, the much greater difficulties of trading electricity, gas, and other carbon dioxide relevant commodities on a black market compared to the less centralised goods of food and clothing, which rely on a quite different distribution infrastructure. Of course, a digital system and more centralised infrastructure also allow for (more sophisticated kinds of) fraud; still, it is less likely to reach the extent that existed in relation to wartime rationing.

252 See, for instance, the rough-and-ready calculation by Müller (2011: 95 ff.).

In order to test the public acceptance of personal carbon trading, various empirical studies have been conducted. If compared to other instruments, such as carbon taxation, there seems to be a preference for personal carbon trading (Fawcett 2010: 6873), which does, however, hinge upon the specific design of each instrument (Bristow et al. 2010). The overall acceptability of personal carbon trading, according to a study, can reach 80% (Bristow et al. 2010: 1833). If, however, the high costs of setting up the system were clearly communicated to survey participants, it is likely that public acceptance would be lower.

Finally, and usually disregarded in the literature on this subject, it should be considered what impact personal carbon trading would have on individual liberty. As an instrument that is explicitly attached to individuals, it might come along with an especially heavy burden on liberty. There are two aspects of this instrument that have the potential to influence the freedom of those concerned by it: the distribution of emission allowances[253] and the mechanisms needed to enforce the system. First, it is an inherent characteristic of personal carbon trading that there is an upper limit up to which individuals can engage in certain activities that lead to carbon dioxide emissions, thus limiting their freedom to further engage in these activities once they have used up their budget. However, this mechanism that limits liberty is alleviated as individuals are free to purchase a higher allowance, provided that they can afford it and that there are others who do not need their full budget. In addition, the instrument leaves individuals open to decide on which of their activities to spend their carbon dioxide emissions, thus restricting their freedom less than any instrument which addresses certain consumption activities. As discussed in chapter 4.2.2, the accumulative harm produced by the sum of individual carbon dioxide emissions has the potential to significantly restrict the liberty of others (living in the future or in poorer regions of the globe), which provides an adequate justification for curtailing the individual freedom to emit carbon dioxide. However, once there is a cap put on overall emissions, it is still questionable how they should be distributed. The distribution pattern of emission allowances has the potential to restrict the liberty

253 Note that most of the following discussion deals with distributive justice rather than liberty, but as said above, the outcome of the distribution is likely to have effects that limit liberty. The relationship between the ideals of liberty and distributive justice is a complicated one, which I cannot adequately cover here.

of some much more than that of others.[254] As said before, proponents of the idea usually suggest an egalitarian distribution of emission rights, which they consider to be the fairest solution. Because of concerns about the distributive impact on low income groups, some modelling studies have been conducted in the UK. Accordingly, an egalitarian allocation of emission rights would rather benefit low-income households as they tend to have lower-than-average emissions and could sell their surplus allowances. However, there are also some low-income groups who would be worse off as they cause higher emissions (Fawcett 2010: 6870 f.). Still, even if it is assumed that most people from low-income groups would benefit from personal carbon trading, this does not answer the question of whether an equal distribution would be just. Even leaving aside the abstract philosophical discussion about distributive justice and equality (see section 2.3.3; cf. Hyams 2009, Starkey 2012 a), there are sufficient empirical aspects that evoke concerns about the justice of an equal distribution of emission rights. For instance, people who live in rural areas or colder climates tend to cause higher emissions, and personal carbon trading might put a rather severe burden on them[255]. Some people spend a lot of time at home because they care for children, work in their home office, are elderly people, ill, students, or unemployed, and thus will need a relatively high amount of energy, for which they would have to surrender more emission rights than people who spend a lot of time at their workplace. How should they be compared fairly? Just because someone does not use a lot of energy for private use does not mean this person does not have a high energy use. A further open question is how children should be considered in the allocation of allowances (Seyfang et al. 2009: 12). In addition, it needs to be discussed whether it would be fair if only the emissions caused by individuals from their use of household energy and transport were covered, as some individuals rely more on these, while others more frequently purchase goods with embedded emissions, which are not included in personal

254 It should also be noted that it is questionable why carbon dioxide emissions should be subject to a certain distribution and not other environmentally harmful emissions or scarce resources, etc., as well. Would fairness not require us to limit and distribute all of them according to a certain principle? This brings us back, however, to the problems already discussed in section 4.3.6 and will thus not be dealt with again here. Cf. also the end of section 5.2.1.

255 It may of course be questioned why people living, for instance, in rural areas, who have chosen to live there themselves, should be allowed to cause more pollution than others.

carbon trading. Emissions caused in the production of goods might be dealt with by including producers in the trading mechanism and requiring them to surrender emission certificates accordingly, which would, however, also have different distributive effects on individuals and thus come at the expense of the chosen distribution mechanism (Hyams 2009: 239).

Thus, the idea of allocating carbon allowances to individuals based on a certain distributive principle seems to be a fair and simple solution initially, but when we look at its practical intricacies many reasons can be found to deviate from the distribution principle in order to safeguard justice. However, as Hyams (2009: 251) has argued, even if any distribution of carbon allowances will always only be an approximation of the ideal, "practical difficulties do not impugn the moral value of the distributive principle itself, and an approximation to an ideal is still a better outcome than complete disregard for the ideal." Given the existing trade-off between practicality and justice here, creating a rather simple system in the beginning might be considered: carbon certificates could be distributed equally first and some certificates could be kept in stock for those who can reasonably claim additional certificates because of unchosen and exceptional circumstances, and the system could be improved over time (Hyams 2009: 252). Still, taking all justified "unchosen exceptional circumstances" into account in the allocation mechanism might over time make personal carbon trading an extremely complicated instrument that requires a significant apparatus of bureaucracy.

Thus, besides questions concerning the impact of the distribution of allowances, surveillance may restrict individual liberty. Even if, as a concession to practicality, only some sources of carbon emissions are concerned and emissions embedded in products were excluded, significant amounts of data about individuals would possibly have to be collected. As opposed to this, in the case of taxes not individuals but rather large producers would be the main target of surveillance. It is thus questionable whether a system of personal carbon trading would not interfere with individual liberty too severely given the surveillance mechanisms required.

5.2.3 Planning instruments for urban transport infrastructure

The final policy example addresses a certain consumption field: individual motorised transport, in particular private cars, is one of the main objects that cause anger among environmentalists. There are obvious reasons for this, as significant environmental harms are associated with cars (OECD 2011 a: 98, Golub 2016: 263 f.): passenger cars, for instance in Europe, cause around 12% of total European carbon dioxide emissions[256], and driving a private car seems to be rather easily avoidable compared to other uses in the transport sector. Motorised road transport has a significant influence on air quality, causing a high proportion of particulate emissions, nitrous oxides, and other pollutants, and is one of the major sources of noise. Cars require space for roads and parking, which is especially a problem in urban environments, but also with regard to land use in general. The production of passenger cars requires significant amounts of materials, the exploitation of which causes further environmental impacts. Not least, cars pose the risk of other traffic participants, such as pedestrians and cyclists, being injured or killed in an accident. A recent study which compared the social costs and benefits of the two most important urban transport modes, the car and the bicycle, considering the aspects accidents, climate change, health, and travel time, found that the costs of car driving are more than six times higher than those of cycling (0.50 versus 0.08 euros per kilometre) (Gössling/Choi 2015). Car driving therefore does not only entail potential harm to others, but is also a rather expensive way for society to get around.

There is a purely technical solution to at least some of these problems which is advocated by many politicians, industry lobbyists, and others: carbon dioxide, other air emissions and, to some degree, also noise could be effectively reduced through the replacement of gasoline with electric motors, in conjunction with a full electricity supply with renewable energies. However, the progress in developing this technology does not appear to be fast enough to replace a significant share of gasoline engines within a reasonable time frame[257], which is why it seems indispensable to apply the strategy of reducing the number of cars and kilometres driven in parallel.

---

256 See http://ec.europa.eu/clima/policies/transport/vehicles/cars/index_en.htm (last accessed December 2015).

257 For instance, the German government aims to increase the fleet of electric vehicles in Germany to 1 million by 2020 (Bundesregierung 2009), given a total vehicle fleet of 62.4 million in 2015, out of which 44.4 million are passenger cars. Despite strong political efforts, (even) this goal is currently out of reach—in early 2015, only around 19,000 electric passenger cars were registered in Germany, even though it should be noted that the share of electric cars is increasing (Kraftfahrt-Bundesamt 2015).

Of course, even sufficiency advocates do not see private cars only as a problem, as they also fulfil many socially important functions (Kemp 2010: 9), and in less densely populated areas there is much less discussion about car use than in urban areas, where they seem to be more dispensable[258]. An important target in the context of sufficiency is thus to reduce the number of privately owned cars, particularly in cities, and achieve a shift to more environmentally sound traffic modes, such as public transport and cycling. Numerous policy instruments are available for this, as Table 9 shows: regulatory, planning, economic, as well as informative measures. The focus here is on planning instruments targeted at changing infrastructure in order to support non-motorised and public transport, as these are comparably effective (see below).

Possible infrastructure measures in the sense of sufficiency include both measures that make individual car use less attractive and the use of other transport modes more attractive: reducing parking space, creating park and ride options, supporting car sharing schemes with specific parking spaces reserved for them, improving conditions for pedestrians, cyclists and users of public transport in cities with dedicated lanes and programming of traffic lights, establishing traffic calming zones and completely car-free areas, setting up bike share programmes, generally improving the options of public transport, supporting the "compact city" of short distances with a mix of living and working areas, and removing incentives to move to sparsely populated areas. Some transport policies are a mixture of planning and other instruments; for example, congestion charges in city centres consist primarily in a financial incentive (or disincentive) and, at the same time, focus on a specifically chosen area; speed limits or bans on car use in certain areas are regulatory instruments that also focus on specific areas, and thus can be considered planning instruments at the same time[259].

The policies mentioned encourage the adoption of transport practices in accordance with sufficiency and make the resource intensive practice of

258 Here again it may be questioned why people in rural areas should be allowed to cause more pollution than others (cf. footnote 255). Nevertheless, in the sense of the approach of addressing environmentally relevant consumption activites, it is appropriate to start with cases where these are more easily replaceable rather than other cases.

259 Cf. also Golub's (2016: 262) differentiation between four approaches of transforming urban mobility systems towards sustainability: price signals, land use changes, technology development, and communication.

individual car driving less attractive. Transport infrastructure as an inherently public issue is a particularly good example of individual motivation to adopt lifestyles in line with sufficiency not being enough. For instance, if it is dangerous and uncomfortable to go to work by bike because no dedicated lanes are available and cars pass by at high speed, and if no acceptable public transport option is available, individuals will very unlikely do without their own car if they can afford one.

As regards the effectiveness of different transport policies, including those related to infrastructure, there are numerous empirical and modelling studies. These indicate that the given transport infrastructure has a clear influence on the choice of urban passenger transport modes. Other important factors that influence the choice of transport modes in cities are urban size and density, socio-economic factors, and the costs of different means of transport (EEA 2013 b: 47 ff.). The share of cycling as a mode of transport for short distances increases with dedicated bicycle infrastructure policies (Rietveld/Daniel 2004, Macmillan et al. 2014). This effect is also influenced by the performance of competing modes of transport relative to bicycles in terms of speed and costs (Rietveld/Daniel 2004). Cities like Copenhagen have experienced a significant increase in the share of cycling of the different modes of transport since the introduction of a cycling infrastructure policy (cf. City of Copenhagen 2015). Generally, more coercive measures, such as prohibiting car use in certain areas or at certain hours, can be expected to be more effective than less coercive measures, but they can have other disadvantages such as feasibility problems (Gärling/Schuitema 2007). Combinations of different measures, for instance a transport infrastructure that is benign towards cyclists, pedestrians and users of public transport combined with financial incentives and information campaigns, are likely to be most effective (Gärling/Schuitema 2007, Cairns et al. 2008)[260]. In any case, any measures adopted should be carefully evaluated, as they sometimes lead to effects contrary to what is desired: a bike share programme in London resulted in a significantly higher amount of kilometres of motor vehicle use because of the truck use required for rebalancing of bicycles (Fishman et al. 2014).

In terms of the efficiency of different transport policy instruments, no studies could be identified that directly compare changes in infrastructure

260 Information campaigns alone, according to another study, cannot be expected to have an impact on transport behaviour and can influence attitudes even negatively (Tertoolen et al. 1998).

with, for instance, economic instruments[261]. Changing infrastructures is certainly accompanied by high costs and requires time, but there are also some planning measures that can, in principle, be implemented immediately and at rather low cost, for instance reducing parking space, establishing traffic calming zones, lowering speed limits, and certain economic incentives linked to selected areas. Besides the costs of setting up a policy, its enforcement and maintenance has to be considered: the compliance with speed limits and no-parking areas needs to be surveilled, and infrastructures entail maintenance costs. However, speed limits, streets and traffic signals are present in modern cities anyway, and after changing them once, maintenance and surveillance costs can be expected to be about the same as before. Generally, it can be expected that policies which aim to encourage the use of other modes of transport are less likely to suffer from implementation deficits and require less enforcement efforts than those which aim to discourage the use of private cars.

An important aspect for changes to urban transport infrastructure is their political feasibility and acceptance. Like personal choices regarding food (cf. chapter 1), choices of means of transport tend to be charged with emotions, and any external influence on these choices is likely to invoke strong reactions. On the other hand, it seems that the share of cycling and walking in European cities is increasing (EEA 2013 b: 46), and there is a trend towards the adoption of specific bicycle strategies and plans in order to increase the modal share of the bicycle as a means of transport. The private car appears to be losing ground as a symbol of independence and status, at least among young people. Car-sharing systems are increasingly used (Shaheen/Cohen 2013) and provide an attractive alternative to individual car ownership. So far this has not yet resulted in an overall decrease in car ownership and kilometres travelled by car[262], but there are trends among young people in wealthy countries that have led to a reduction in car ownership and driving licence availability (Kuhnimhof et a. 2012). Measures that tend to arouse public opposition, such as congestion charges

---

261 One research article titled *Comparing the efficiency of alternative policies for reducing traffic congestion* (Parry 2002) focuses on economic instruments only.

262 According to Eurostat data, car ownership in the EU has been increasing steadily even since the year 2000, and even in most countries with an already high level of car ownership; the share of kilometres travelled by passenger car per person has stagnated or increased in most EU countries since 1990 (see http://ec.europa.eu/eurostat/web/transport/data/main-tables; last accessed October 2015).

in city areas, achieve greater acceptance if they are both linked to information measures that communicate their benefits to the public and if they prove to be effective in reducing the problems (congestion) they aim to reduce (Gärling/Schuitema 2007).

As regards the impact of planning instruments for transport infrastructure on individual liberty, it should be pointed out first that taking part in traffic in a densely populated area is as such a social activity that has to be based on rules which need to be determined at a political level, because without rules it would be very dangerous for everyone. Thus, in mainland Europe we drive and cycle on the right side of the street, and few people would see this as infringement of their liberty. Improving the conditions of those who participate in traffic with a certain means of transport often comes at the expense of those who use another means of transport: for example, building more bike lanes decreases the available space for cars or pedestrians. If someone complains that a certain policy is attacking his or her "liberty as a car driver", the decrease in liberty they feel is coupled with some change to the status quo, that is, public space is redistributed to the benefit of other traffic participants. The fairness of the status quo is usually not called into question, and the improvement in liberty or security for other traffic participants is not considered. Referring to principles of distributive justice, however, appears to make little sense with regard to transport infrastructure for different transport modes. Instead, as described at the beginning of this section, it is precisely the intention of sufficiency policy to reduce the modal share of one means of transport, namely the private car, because of the (potential) harm to others entailed in car driving. Thus, arguments in favour of car-driving based on individual liberty can be countered easily by the harm principle, as there are many aspects of car driving that cause potential harms to others, including both accumulative (e.g. emissions) and individual harms (accidents), and there are less harmful alternatives available.

5.2.4 Conclusions from the policy examples

Three policy instruments that could play a major role in establishing a policy of sufficiency were analysed in this section. The first two, environmental taxes and personal carbon trading, are both economic instruments and are therefore similar in some respect. At the same time, they are different in terms of which point they are attached to: taxes focus on the resource or pollutant targeted and thus on the ideal object in the view of economic theory, while personal carbon trading focuses on the individual and thus represents the perspective of justice theory. The third example regarding transport infrastructure is from the category of planning instruments and focuses on a specific consumption area (cf. section 4.3.6 on this distinction).

These instruments are not presented as either-or options here, but rather as examples from a vast array of instruments and could also be implemented in parallel. A stand-alone policy that would be the ideal way to implement sufficiency as policy is not in sight. As opposed to what theorists sometimes suggest, for environmental taxes and personal carbon trading in practice only a partial and non-ideal implementation of these instruments is possible, which takes its toll on their assumed efficiency and fairness, with the result that they ultimately come close to the pragmatic solution of regulating selected consumption activities. For instance, if a tax was imposed on greenhouse gases, in practice this would mean that the most relevant and easily surveillable sources of greenhouse gases would be taxed[263] and that there might be certain exemptions for economic or other reasons. More severe appear to be the problems related to personal carbon trading, which is why I am rather sceptical about the feasibility of this instrument. As opposed to personal carbon trading, the other two policies are already widely practised. However, in order to lead to significant reductions in environmentally harmful consumption, they would have to be implemented more consequently, that is, by applying stricter tax rates and more stringent changes to infrastructure in order to reduce individual car use.

In terms of them supporting and encouraging individuals to adopt behaviour in line with sufficiency, all three instruments have certain advantages: environmental taxes change economic incentives and thus reward environmentally sound behaviour; personal carbon trading is a practical application of the idea of environmental space (cf. section 4.3.6) and makes individual emissions more visible, thus sensitising people to this issue and allowing them to compare with others; creating an infrastructure that favours environmentally sound modes of transport changes the framework conditions that are important determinants of individual behaviour.

263 That is, the combustion of fossil fuels and maybe cattle breeding, but probably not land use related sources or the combustion of biomass, and very unlikely emissions "embedded" in other goods.

With regard to the other evaluation criteria that have been used—effectiveness, efficiency, likelihood of implementation deficits, political feasibility, as well as the impact on individual liberty—the instruments discussed are, at least partly, too different to allow for a direct comparison. While the intention of the preceding discussion was just to show the advantages and problems related to some examples of sufficiency as policy, a more systematic comparison of all policy options that exist in a certain case (e.g. reduction of carbon dioxide emissions) would be needed in order to find the most suitable option or combination of policies.

Most likely, if there was a political intention to really *reduce* environmentally harmful consumption, what would be needed is a mix of different policies that, first, are compelling enough in terms of both regulation and setting significant economic incentives while involving as little arbitrariness as possible; second, that provide framework conditions which enable and further incentivise (by non-economic means) consumption patterns in line with sufficiency; and finally, that should also include ancillary tools for information and communication purposes.

5.3 A "political conception" of sufficiency

The preceding chapters discussed the background to and rationale behind why sufficiency should be adopted as a strategy not only for individuals, but also on the political level. Potential approaches as well as according difficulties were presented. In this section, I intend to propose a "political conception"[264] of sufficiency. This conception is embedded within the framework and based on the arguments presented in the preceding chapters and consists of a description of its normative background, a definition and more concrete proposals as regards the adoption of policies. Later on in this section, I will discuss its limitations.

---

264 On the distinction between "concept" and "conception", cf. chapter 1, footnote 3.

5.3.1 Conception

Area of application and normative basis

The conception is embedded into a framework of theory strands as presented in chapter 2: it applies only to liberal-democratic societies, which are assumed to strive for at least a specific minimum conception of sustainability that is in line with their inherent requirements, as it serves to protect certain rights and liberties and is needed to preserve the livelihood these societies are based on. The liberal component is characterised by individuals' equality of moral status, by liberty from external interference in one's own affairs except where it is limited by others' liberties and rights, in particular by the harm principle, by neutrality towards individual conceptions of the good, and represents a political rather than an economic variant of liberalism (section 2.1). The democratic component features a collective decision-making procedure that is based on equal voting rights and characterised by deliberation among the participants, which consists in an exchange of justifications they give each other to support their respective positions (section 2.2). Taken together, the liberal and the democratic components build a complementary framework, both safeguarding individual rights and liberties and requiring those in power to respect the will of the majority.

Sustainability is proposed here as a rather uncontroversial minimum approach, meaning that it does not exclude per se more "ambitious" approaches, but is restricted to this basic version for the purposes here. Even if characterised as a minimum approach, this, however, does not mean that it is simple or uncontroversial to find ways to achieve the aims of sustainability in practice. In this sense, sustainability comprises the following elements (section 2.3): anthropocentrism, a justice conception that focuses on meeting the needs of contemporaries and future generations rather than striving for distributive equality, a focus on sustaining the "critical" stocks of natural capital (or: environmental carrying capacity) rather than preserving the sum of natural capital or capital in general, and indifference regarding the question of economic growth as long as this does not jeopardise the main aim of sustaining the desirable state characterised by the previous elements.

Further, the conception proposed here is limited to wealthy countries, where typical consumption patterns are clearly "excessive" in the sense that their aggregated environmental effects are harmful and where average individual consumption by far exceeds what would be sustainable if all people worldwide consumed at this level (cf. section 4.3.3).

Characterisation of sufficiency as policy

While technical solutions are important to achieve the aims of sustainability, it is indispensable to realise the sufficiency strategy as well, because technical solutions alone cannot effectively stop the most pressing environmental problems (chapter 3).

Sufficiency is defined as "the reduction of demand or use of goods and services with high environmental impacts in order to achieve per-capita consumption levels that ensure emissions and resource use to stay within the environmental carrying capacity" (section 3.2.3). Sufficiency in this sense is considered a neutral necessity rather than a good or bad thing. By us linking it to consumption levels in line with the Earth's carrying capacity, it acquires an absolute dimension, which indicates that it does not only aim at relative reduction in relation to something else (e.g. GDP). Further, reference to goods and services that are associated with relatively high environmental impacts clarifies that only these are the focus of sufficiency, and not consumption per se.

This definition does not allow us to clearly distinguish it from technical strategies aiming to limit adverse environmental impacts, and there may be overlaps with these. It is indeed very difficult to separate these two approaches clearly from each other (section 3.2.2). Nevertheless, it can still be said that while sufficiency has an emphasis on behaviour changes and the consumption side, technical solutions primarily address products and production processes (cf. Table 4 in section 3.2.2). Overlaps with technical solutions are, however, not considered a problem here as both kinds of strategies are seen as complementary.

Further, sufficiency needs to be adopted not only on an individual but also on a "political" level, as individual practices, even though important, are not effective enough alone (section 4.1). Accordingly, *sufficiency policy* includes policies that, openly or implicitly, aim for sufficiency in the sense described above. What is crucial in the notion of "policy" or "political" is that political action targets the collective level and aims to bring about collectively binding decisions (see introduction to chapter 4). In my discussion here, this only includes activities of the state (on its various administrative levels, including international commitments) and not other actors' activities aimed at the political level (e.g. by non-governmental organisations). The main reason for this is that I intend to provide a justification for sufficiency measures taken by the state, and this justification is not needed with regard to the activities of other actors. Thus, the exclusion of the latter is not meant to show disregard, and it is acknowledged that non-state actors are, of course, important and can be a precursor to action by the state.

Justification

In line with the focus on environmental carrying capacity, sufficiency policies are justified through concerns about the harmful cumulative environmental effects of certain consumption patterns rather than, for example, controversial claims of what constitutes the good life or paternalistic reasoning. The underlying rationale, the claim contained in the harm principle that society may intervene if the actions of someone involve harm or significant risk of harm to others, is widely accepted, especially among liberals (section 4.2.1). Likewise, a case can be made that society has the right to interfere in individual action in instances of accumulative harm, that is, harm that is caused by the sum of many small, uncoordinated contributions (section 4.2.2). It should be noted that as a consequence of choosing an anthropocentric approach (cf. section 2.3.2), the notion of harm applies only to human beings under this conception. The reason for using the harm principle is that policies aiming at sufficiency can have significant impacts on individual freedom; therefore, such policies should be based on the strongest possible rationale, causing, as such, little disagreement (disagreement, however, is likely to arise in case-by-case discussions about the severity and likelihood of harms and the role of consumption in the causality chains that lead to these harms; cf. Box 2 in section 2.3.4, and section 4.3).

Hence, the aim of sufficiency as policy is to influence individual consumption patterns in such a way that aggregated consumption stays within the limits of the Earth's environmental carrying capacity and to avoid the thresholds beyond which there would be harmful environmental change being exceeded (on a global, regional, or local level). In other words, sufficiency policy aims to sustain the "critical" natural capital needed to preserve human livelihoods (cf. section 2.3.4).

The same rationale serves to justify most environmental policies, also policies which promote technical solutions (cf. section 4.2.2). However, since the latter tend to involve less significant interference in individual liberty than policies that call into question consumption levels, they are usually less controversial (cf. section 4.3.2)[265].

Environmental carrying capacity and critical limits

Consequently, sufficiency policy should first of all be informed by the limits of environmental carrying capacity. These limits are defined by the level of emissions to environmental sinks and the level of use of natural resources that characterises the point beyond which there would be harmful consequences to human beings now and/or in the future and they could no longer meet their needs. The limits are located on different levels corresponding to the kind of environmental problem, including the global (e.g. for greenhouse gases), regional (e.g. exploitation of fish stocks), and local levels (e.g. certain types of water and air pollution). Of course, it may be very difficult to quantify such limits in practice, as it is controversial what elements of the environment are critical to human livelihoods and at which levels limits should be fixed in order to take into account risks and uncertainties (section 2.3.4). However, there is a high certainty with regard to the essentiality of some environmental functions for human beings (e.g. the functions provided by the ozone layer, freshwater reserves and a stable climate system), and limits can be defined based on scientific findings at least for these functions, even if they may have to be corrected later. With a view to the "minimal" conception of justice used here, it follows that at least a level of critical natural capital should be safeguarded that allows human beings to meet their basic needs and not be harmed. If there is uncertainty about the critical limits of pollution or resource use and if significant risks are involved, this requires according deliberation procedures to find an appropriate level of precaution. It is thus assumed here that it is, in

265 A different reason why technical solutions are less controversial is that sufficiency, as it aims at quantitative restrictions, is much more likely to curb economic growth. However, as economic growth is not a precept of sustainability or liberalism (sections 2.1.4, 2.3.5), this does not require independent justification, but rather anticipation of possible practical consequences (cf. section 5.3.2; cf. also footnote 208).

principle, possible to determine such critical limits (and there are attempts in this sense, cf. section 2.3.4), even if scientific controversy continues and the environmental functions in question are too complex to allow for the settlement of such a controversy.

Further, it should be noted that it appears to be more difficult to establish a link between overconsumption of certain resources and harm to others in comparison to the overuse of environmental sinks and such harm (cf. section 4.3.4). While polluting the sinks produces harm more directly (even if causality chains are quite complex in some cases), potential harmful effects because of the overexploitation of resources seem to be more indirect due to a higher potential for substitution and globalised resource markets, where manifold influences on resource availability exist. Still, it should be noted that the practices used to exploit natural resources often produce very harmful emissions to soil, air, and water, too.

Practical suggestions for sufficiency policies

The harm principle itself does not contain advice on how harm should be stopped. Unlike in individual harm cases, for accumulative harm it is usually not possible to impute the responsibility for the harm directly to individuals (cf. section 4.2.2). Thus, the principle gives a reason to the group that is causing (likely) accumulative harm to organise itself in such a way that the harm is stopped.

As an absolute notion (cf. section 3.2.2), sufficiency requires agreement on absolute levels of adverse environmental impacts, which should not be surpassed in order to prevent accumulative environmental harms. Such targets that define the (global, regional etc.) sustainable macroeconomic use of sinks and sources can be established based on information about the environmental carrying capacity, as is already done, for instance, for greenhouse gases (and other impacts, cf. section 2.3.4 on planetary boundaries). They can be further broken down to states or other administrative units by collecting information on an administrative unit's contribution to the problem. If, as a rule of thumb, the average per-capita contribution within an administrative unit is higher than what would be sustainable if everyone (who lives in the area relevant for the problem) contributed to the problem at this level and the sum of contributions exceeded or may in the not so distant future exceed the sustainable level (i.e., consumption is "excessive", cf. section 4.3.3), the administrative unit should adopt policies that are effective in curbing its overall contribution sufficiently[266].

Three idealised ways to translate limits on "excessive" aggregate consumption into policies that lead to a reduction in individual consumption can be distinguished between (cf. section 4.3.6): to directly address certain resources and sinks (e.g. by means of taxation or exploitation quotas), to put a cap on individual contributions (by establishing individual pollution or resource use allowances), and to focus on certain consumption activities that are environmentally harmful. While the first approach supports individual sufficiency by correcting economic incentives and aims to internalise external costs, the second one makes the limits for a consumption pattern in line with sufficiency visible by establishing an "environmental space" that individuals can make use of, and the third one supports individuals by making it easier for them to adopt certain practices in line with sufficiency or making it less attractive not to do so. The first two approaches, derived from economic theory and justice theory respectively, are sometimes praised as perfect stand-alone solutions. However, if they were implemented consequently in order to effectively reduce individual consumption, they would be likely to produce quite strong undesirable side effects such as extensive surveillance needs (cf. sections 4.3.6 and 5.2). Thus, even if, for instance, taxes as one instrument representing the economic approach are likely to play a major role within a sufficiency policy, they should be supplemented with the third approach, which targets certain consumption activities (e.g. regulating traffic in the sense of sufficiency, banning certain goods, or "nudging" people into certain behaviour). This approach, which is a piecemeal approach rather than one based on theoretical reasoning, is not ideal in terms of efficiency and fairness, as some will always be affected more than others. On the other hand, it is, depending on the instruments chosen, more likely to be accepted than, for example, a significant increase in taxes and is also able to mitigate socially undesirable side effects.

266 Here, this is just to be understood as a simple rule to acknowledge that there is a problem and something needs to be done. Which countries should reduce their contributions to an accumulative problem, by how much and based on which principles, for example in the discourse on climate change, is, of course, a highly contested question in practice (cf. e.g. Caney 2009, Miller 2008). Cf. also the discussion on equality as a distribution principle (section 2.3.3).

Following the agreement on environmental targets and these more general considerations about possible approaches to decreasing consumption, it should be considered which concrete instruments of environmental policy are suitable to realise sufficiency as policy. Generally, they can be chosen from the full spectrum of environmental policy instruments ranging from more behaviour-determining to more information-based and voluntary approaches: regulatory law, planning, economic instruments, nudging, cooperation, and information (section 5.1). Sufficiency policies do not necessarily have to be new or additional instruments, but will often consist in the more consequent application of existing environmental policies. Within a policy of sufficiency it seems to make sense to assign a major role to regulatory, planning, and economic instruments rather than the remaining "softer" categories, because cooperation and information rely again on individual adoption of the behaviour desired (which is important but deemed insufficient though, cf. section 4.1.1), and with regard to "nudging", the scope for application appears to be limited in this realm (see section 5.1.3). At the same time, information, cooperation and "nudging" can be supportive as ancillary tools and help to increase the acceptance of sufficiency policies. If there was the political intention and public acceptance of reducing environmentally harmful consumption, a mix of different policies would be advisable, which would in part consist in amendments to existing policies (i.e. more stringent taxation, more consequent traffic policies that favour environmentally sound modes of transport, etc.). Taken together, these should be compelling enough, involve as little arbitrariness as possible and if necessary contain mechanisms to mitigate negative (social) side effects, enable people to adopt behaviour that relies less on material consumption, and also include ancillary tools for information and communication purposes.

In the choice of instruments, criteria for the evaluation of environmental policy instruments such as effectiveness, efficiency, the likelihood of implementation deficits, political feasibility, and the impact on individual liberty should be taken into account. With regard to sufficiency policy, the latter two criteria are of particular interest. First, as regards political feasibility and acceptance, it is an empirical question whether sufficiency as policy and which particular policies will be accepted by a majority before certain environmental problems and scarcities become tangible in wealthy, democratic societies. As the history of environmental policy shows, the prospects of a measure to be supported by a majority can change over time and are also influenced by external events that create opportunity windows. In any case, good arguments on the side of those in favour of sufficiency policy will be supportive in communication and deliberation processes (cf. section 4.4.4). Second, sufficiency policies tend, compared to technical solutions, to interfere more significantly with individual liberty (section 4.3.2)—at least those that do not only inform, encourage and enable individuals to adopt less resource-intensive consumption practices. According to the principle of commensurability, only measures that are proportionate and effective may be adopted, and among those which satisfy these two criteria, the measure should be selected that has the least significant adverse impacts on those concerned as well as the general public (cf. section 4.3.5). Thus, if no technical solutions with acceptable side effects are available within a reasonable time and at a reasonable cost, and if encouraging and facilitating sufficiency is not effective enough, more compulsory policies that require sufficiency practices are also in principle legitimate in order to bring the average individual contribution to a sustainable level.

As the aim of sufficiency policy is not to condemn and curb consumption as such, but only to decrease environmentally relevant consumption, it should concentrate on the consumption fields that have been identified as environmentally most relevant, namely the mobility or transport sector, food, housing, and electrical household appliances. An open question is how the environmental impacts related to imported goods and their production chains should be dealt with, which, from a consumption perspective, should in principle be allocated to the consumer or to the importing entity. However, the complete collection of according information can be nearly impossible or at least very expensive, as production chains are often long and complex.

A further note concerns the administrative level on which a political solution to a problem should be sought. Ideally this should be done according to the character of the problem to ensure effectiveness, that is, a global solution to a global problem etc., which will, however, not always be feasible. For instance, common taxation schemes are already a problem at European Union level. Typically, problems of global character are dealt with in the international arena rather by means of agreements that contain common targets, which need to be fleshed out by the individual signatories to the agreement with concrete policies on national level. If it is not possible for the moment to find a solution on the level that corresponds to the kind of problem in question and not all parties concerned are disposed to participate, one might ask whether individual states or smaller administrative units should still adopt such policies—the usefulness of national approaches to sufficiency policy might be questioned in a similar way as individual strategies. However and also similar to individual strategies, policies on lower administrative levels are a precondition for establishing good practice examples that can be followed by others, and increase the credibility of affluent nations in global negotiations.

5.3.2 Open questions and limitations of the conception

The conception presented above provides a pragmatic approach to realising sufficiency on political level within liberal-democratic societies. However, its potential success in effectively reducing the external effects of certain consumption habits and leading to sufficiency is also influenced by some aspects that could not all be analysed in detail in the preceding chapters, as this would have partly gone far beyond the scope of this study.

First of all, there are three aspects that I have completely excluded from my analysis. These are population size, the question of whether sufficiency also applies to less wealthy or even poor countries, and what non liberal-democratic societies should do in terms of sufficiency policy. As regards the first one, there are some writers who do not only discuss the need to change consumption patterns under the headline of sufficiency, but also a need to limit population size (cf. section 3.2.2). While it may be true that this is equally as relevant as consumption patterns, I think this issue has even greater potential to meet with refusal, and it is currently also controversial whether regulating population numbers is needed at all. Even if it is true that liberal theory "is not a suicide pact", and "within the context of a certain catastrophe that threatened the human race, reproductive choice surely could be regulated" (Coglianese 1998: 57), I have preferred to exclude this thorny issue from my analysis.

Similarly awkward is the question of whether sufficiency should apply to societies that are not yet rich, too, given that developing countries account for a relevant and increasing absolute share of environmental impacts, for instance around 60% of global carbon dioxide emissions (in 2012; UN 2015 a: 53), and given that macroeconomic rebound effects from potential sufficiency policies elsewhere might lead to a further increase in the absolute environmental impacts in developing countries (cf. section 3.2.4). However, I think that technical solutions alone can still do a lot in these countries to curb adverse environmental effects, and without bringing consumption patterns in line with sufficiency within wealthy countries, which have caused an impressive cumulative historical burden on the environment, it is hardly possible to ask poorer countries not to catch up to a somewhat higher level of consumption.

Finally, there remains the question of what non liberal-democratic societies should do in terms of sufficiency policy. While most wealthy countries have liberal-democratic governments, there are also some that do not, and the consumption patterns of which are certainly environmentally significant. I must admit that I have no answer to this, as the reasoning presented here makes sense only within liberalism.

Further, there are two other aspects that are more relevant for the countries focused on here, which I have not fully analysed but only briefly touched upon in previous chapters. This includes, first, the potential consequences for the economy and the labour market and the related social consequences if such a policy was realised (cf. section 4.1.2). While I have stated that, in my opinion, sustainability is just not concerned with the development of GDP (cf. section 2.3.5), economic development in general does, of course, have an influence on the development of a society and the well-being of its citizens. In theory, it is possible that widespread adoption of sufficiency practices would mainly shift economic prospects from resource-intensive to labour-intensive sectors; however, such a restructuring of the economy would already have at least temporary negative consequences and produce considerable uncertainties for those who work in resource-intensive sectors, which would probably trigger cascade effects. It should thus be noted that sufficiency policy would have (temporary) adverse effects on the economy, which may be significant and thus require further analysis.

Second, I have only briefly mentioned the limitations of the approach used here with regard to the insights from political steering and governance theory (cf. section 4.1.2). A policy or political programme is not to be mistaken for a mechanical device which will carry out its tasks and produce the desired results as soon as the "on" button is pressed. Instead, there is always a chance that a policy is not implemented and enforced properly, that those concerned will oppose its implementation or find ways to circumvent it, or that the global and transnational character of an environmental problem will counteract the effects of a national policy. Such a sensitive and complex area as consumption might be particularly difficult to "steer" via policies. There are already many policies in place that aim to influence individual consumption habits in one way or another (consider the policies with regard to smoking, "unhealthy" food, or "sustainable consumption", cf. section 3.2.5), and it might be useful to analyse their effects in terms of political steering to anticipate the prospects of success of a sufficiency policy, which could be a subject for future research.

Another limitation from this realm is that my analysis has focused on potential measures taken by the state, thus representing a rather traditional view of "top-down" steering. The involvement of other, non-governmental actors has been excluded from the discussion here. This exclusion is, as mentioned above, mainly useful for analytical reasons. In fact, non-governmental organisations, for instance, can play an important role in the deliberation processes needed to convince people to allow for the adoption of suffciency policies and to adopt individual sufficiency practices. In line with this, non-governmental cooperative solutions to "commons" problems in the sense described by Ostrom (1990) have also not been considered here as it is assumed that these would not be sufficiently effective for the area of consumption, which, in most cases, involves a far greater number of parties concerned than the classical case of the pasture, and which is, in addition, not limited geographically.

6. Summary and conclusions

The aim of this study was to analyse whether it is justified or even necessary for environmental reasons to adopt sufficiency as a political strategy within wealthy liberal-democratic societies, and if so, to develop a conception of the possibilities and limitations of sufficiency as policy. In order to do so, the preceding chapters picked up the threads of various debates from different disciplines, which all play a role in relation to the subject of sufficiency and policy, to tie them together. To name just the most important ones, the debates introduced here and discussed in the light of the aforementioned research question include the possibility of defining environmental carrying capacity, conceptions of distributive justice, the role of economic growth and of technology within sustainability, the meaning of the idea of sufficiency and obstacles to it, the potential and limitations of individual behaviour as a solution to environmental problems, potential justifications for state interference to protect the environment, the concept of accumulative harm, the role of neutrality within liberalism, the meaning of deliberation in democracy, as well as advantages and disadvantages of specific policy instruments and instrument categories in the environmental realm.

To summarise in more detail, chapter 1 introduced the subject of this study and presented a selection of environmental problems and their (potential) harmful consequences, including greenhouse gas emissions and climate change, the use of natural resources, and biodiversity loss. For these examples of environmental problems, the current trend is not sustainable, and technical solutions alone do not appear to be a sufficiently effective answer, which is why greater efforts are required to get them under control.

In chapter 2, the basic framework for the following analysis was elaborated. The scope of application of the discussion in this study was limited to societies with a liberal-democratic constitution that strive for a certain minimum version of sustainability. It was broadly outlined—for the purposes of this study—what characterises a liberal society and democratic decision-making system: in its liberal component, it is characterised by the equality of individuals' moral status, liberty from external interference in one's own affairs that is limited by the rights and liberties of others, in particular through the harm principle, by neutrality towards individual conceptions of the good, and by representing a political rather than economic variant of liberalism. In its democratic component the system features a collective decision-making procedure that is based on equal voting rights and characterised by deliberation among the participants, which consists in an exchange of justifications they give each other to support their respective positions. Taken together, the liberal and the democratic component build a complementary framework which both safeguards individual rights and liberties and requires those in power to respect the will of the majority. The demands of this framework constitute, on the one hand, a double boundary that political action taken in the name of sustainability needs to respect. On the other hand, it forms the basis for demands of sustainability and can be a driver for these, if, for instance, the protection of certain liberties and rights of some individuals requires the introduction of environmental policies. Sustainability—again, for the purposes of this study—was characterised by the following elements: a pragmatic and enlightened version of anthropocentrism, a justice conception that focuses on meeting the needs of contemporaries and future generations rather than striving for distributive equality, a focus on sustaining the "critical" stocks of natural capital or, in other words, on preserving the environmental carrying capacity, and by indifference regarding the question of economic growth as long as this does not jeopardise the main aim of sustaining the desirable state characterised by the previous elements. The general strategy of chapter 2 was to adopt rather common positions that are compatible with mainstream theorising or constitute a minimal consensus of a whole range of theories, in order to form a maximally robust fundament for the argumentation in the subsequent chapters.

Chapter 3 presented and analysed sufficiency and technical solutions as sustainability strategies. As regards terminology, both concepts are not easy to define, and in particular the term sufficiency has countless interpretations in the literature. The chapter did not come up with a clear demarcation of sufficiency as distinct from technical solutions, but a rather broad definition that allows for a rough distinction and refers to the reduction of demand or use of goods and services with high environmental impacts. It was also shown that many environmentalists, at least in the past, often tended to fiercely advocate one of the strategies, while denying the problem-solving capacity of the other. The analysis here, in contrast, defended the position that both strategies are needed, as both can, in principle, be effective but also have their limitations: for technology, limitations include in particular the fact that no technology can be extended in its application infinitely, as there are conflicts with other societal aims as well as physical limits, for example in the form of suitable space; that all technologies have side effects, including environmental impacts, which also limits public acceptance of their application; that there are problems that cannot be solved by means of technology ("commons" problems); and that the development and diffusion of respective technologies is expensive and requires time. Further, the effectiveness of both technical and sufficiency strategies is hampered by rebound effects, although to a lesser degree in the case of sufficiency. A major limitation to sufficiency are problems of acceptance, because it can have more significant impacts on everyday habits and because there is uncertainty about its effects on the economy and the labour market, if it was adopted by a relevant number of people. Finally, an additional weakness of sufficiency is that considerable time would pass for the strategy to unfold its relevant effects until the required change in values has happened if the strategy is based mainly on individual, voluntary adoption of according practices.

In chapter 4, therefore, sufficiency was discussed as a political approach, besides it being an individual strategy. While the acts of individuals or individual households in the context of sufficiency are considered to be important, already because the individual and societal levels are interdependent, sufficiency as an individual strategy is not effective enough to bring about the required changes in consumption patterns. One of the reasons for this is, among other things, that individual behaviour is embedded into and constrained by the existing systems of provision. It was thus argued that respective action should also be taken on a political level. The "political" was outlined rather narrowly as a notion that refers to actions by the state on its various levels, resulting in decisions that are collectively binding. It was pointed out that the literature on this subject already features a number of publications dealing with sufficiency as policy, but these mostly take a rather practical perspective and lack a detailed analysis of the justification for and objections to such policies. In search of such a justification, the liberal principle that society may intervene when there is an instance or a risk of harm to others ("harm principle") was discussed in detail. A case can be made for the harm principle as an adequate justification for most policies that aim to protect the environment, also when the harm is the outcome of many uncoordinated individual acts ("accumulative harm"). It is, however, more difficult to find appropriate ways to deal with accumulative harm cases compared to harm from a single source, since the causal chain for the harm is not exactly traceable and such harm will often only materialise with a certain probability, and given that the activities that lead to the numerous individual contributions are often beneficial. However, to argue that the costs of avoiding such harms are too high does not hold as a general excuse. In any case, an instance or risk of harm does not create an automatism that a regulation must be set in place, but it does open the discussion about possible measures to deal with the harm. As soon as it is sufficiently clear that the parallel actions of many individuals lead to an accumulative problem that brings about harm, this gives a reason to the community to organise itself in such a way that the activities contributing to the harm are reduced so that the threshold of harm is not reached. This reasoning, in principle, also holds as a justification for policies that aim to limit environmentally harmful consumption. However, it was pointed out that finding ways to stop accumulative harms due to consumption levels is particularly difficult, because compared to technical measures, sufficiency policies—at least those which do not only inform, encourage and enable individuals to adopt certain consumption practices —tend to entail deeper infringements of individual liberty as they aim at quantitative restrictions, while technical policies "only" aim at improving or changing products and processes. The harm principle itself does not provide an answer to the question of how society should intervene in cases of harm to others. Chapter 4 finally anticipated a number of possible critical perspectives with regard to the proposed approach to sufficiency as policy, including potential fears of violation of the neutrality principle, fears that interference in consumption levels might put things at risk that are too important to give up, the view that the harm principle might be too defensive a basis for meaningful environmental protection measures, and the view that sufficiency policy might not at all be feasible in a democracy. According answers defending the approach suggested here were provided to these critical perspectives.

Finally, chapter 5 suggested specific policy instruments that might be suitable in implementing sufficiency as policy. Sorted according to the policy categories of regulatory law, planning, economic instruments, "nudging", cooperation, and information, the instruments suggested are spread across the most environmentally relevant consumption fields (mobility, food, housing, and electric appliances). Special attention was given to three policy examples, which are potentially very broad in their effects: environmental taxes, personal carbon trading, and planning instruments for urban infrastructure to reduce individual motorised transport. As can be learned from these examples, there does not appear to be a particular policy instrument especially well-suited to implementing sufficiency; instead, it is most likely that sufficiency would have to be dealt with on a policy level within a policy mix, often in the form of amendments to existing policies (i.e. more stringent taxation, more consequent traffic policies that favour environmentally sound modes of transport, etc.). Chapter 5 further presented the distillate of the previous chapters in the form of a conception of sufficiency as policy, as well as its limitations and questions that have been left open. Questions that have been omitted include whether population size should also be subject to political interference, whether sufficiency also applies to less wealthy or even poor countries, and what non liberal-democratic societies should do in terms of sufficiency policy. Further aspects that have been mentioned but not analysed in detail and which would require further research and debate include the potential consequences of sufficiency policy for the economy and the labour market, as well as the role of non-state actors and the problems related to political steering.

Taken together, chapters 1, 2, 3 and 4 can be understood as presenting a chain of arguments that leads to the conclusion that sufficiency also needs to be adopted as a political strategy, while chapter 5 analyses ways to do this. Again, in a nutshell, this chain is as follows:

- The current trend is not sustainable with regard to several environmental parameters.
- Liberal-democratic societies should strive at least for a minimum version of sustainability that seeks to preserve the livelihood and certain rights and liberties of people within these societies and worldwide.
- Technical solutions are not enough to deal with the environmental problems we face; therefore, solutions in the sense of sufficiency are needed as well.
- Practising such solutions only at an individual level is not effective enough; therefore, a "sufficiency policy" needs to be adopted on a collective level as well.
- This is also justified, or even required, within a liberal society: the typical consumption patterns of people living in wealthy societies cause accumulative environmental effects, which (are likely to) lead to harm to others living presently and in the future. An action that causes harm to others, including in cases of accumulative harm, gives society a reason to interfere in individual liberty to avoid the harm.

- Regulating consumption levels has a higher potential of the state interfering significantly with individual liberty than promoting technical solutions; therefore, more care is needed in finding appropriate solutions.

The analysis presented here was conducted in rather generic terms and not linked to a particular environmental issue. The aim was to provide a general line of argumentation in view of certain consumption patterns and their related environmental harms, while it is still necessary to see on a case-by-case basis whether sufficiency as policy is an appropriate answer to a problem. Discussing the same issue strictly with regard to a certain example would allow for a more detailed analysis, including the causal chains that lead to environmental harm, according responsibilities, the potential of both technical solutions and sufficiency, and considerations with regard to a just distribution of burdens.

The reasoning on which my conception of sufficiency as policy is based explicitly represents the "lowest common denominator" of a range of convictions and strands of theory. It was constructed as uncontroversially and robustly as possible on purpose in order to allow for the related arguments to be accessible to the greatest possible number of people. Thus, for the purposes here, I did *not* assume a different political or economic system, adopt a physiocentric perspective, or base my arguments on an egalitarian conception of justice or on a specific account of the good life. This may seem not very ambitious from an environmental point of view, from a perspective that highlights the intrinsic and social value of equality, or from a point of view that seeks to emphasise the more positive messages linked to lifestyles that focus less on material assets and more on leisure and enjoyment. However, I consider my approach the most promising one given that there are fears and accusations that sufficiency as policy would end up in totalitarian rule ("eco-dictatorship"). Contrary to such accusations, there is nothing totalitarian in establishing policies based on the harm principle. Instead, they are necessary to safeguard certain rights and liberties of future generations and those currently living who already feel the environmental pressures linked, above all, to the consumption habits of a minority of wealthy consumers. Thus, a liberal policy of sufficiency does not prescribe what consumption patterns should look like, but aims to impede or decrease in relevance certain consumption practices in order to avert harm in current and future generations. As opposed to such formal coercion by means of the law, there would be totalitarian tendencies in attempts to form public pressure to condemn individually harmless acts, requesting or indoctrinating citizens to live certain lives "voluntarily" and making those who do not conform to this lifestyle societal outcasts. Despite my insistence on the harm principle, political measures with the purpose of protecting nature that are less significant in their interference with individual liberty can still be justified in other ways. In addition, the more positive messages of "green" consumption patterns linked to the good life can be used in communication along with political sufficiency measures.

Of course, in a democracy it is necessary that a majority of citizens and/or representatives in parliament support or at least accept the policies that are adopted, and even the soundest reasoning that it is right to do something cannot as such guarantee that it is done. I have to admit that I am not particularly optimistic that policies that lead to a decrease in consumption levels will be adopted systematically in wealthy countries in the near future, given the potential effects on the economy, the existing infrastructures that have been created with significant expenditure, and the (still) prevalent belief that possessing more material assets is generally better than possessing fewer. However, values and attitudes can change over time, and as political decisions are subject to public deliberation in liberal democracies, whether they can convince a sufficienctly great number of people depends on the quality of the arguments brought forward by those advocating a sufficiency policy. As stated before, this is an empirical question, and there are at least gentle tendencies in some wealthy countries that indicate an increasing interest in lower consumption lifestyles, which could pave the way for according policies. Also, political bodies on different administrative levels have recently started to embrace concepts that are linked to sufficiency, for instance in strategy papers on the "collaborative" (or "sharing") economy or the role of social innovations within a policy of sustainable consumption (cf. section 3.2.5; cf. Bundesregierung 2016, EC 2016), even though these papers do not have the explicit intention of really *reducing* consumption. On a communal level, there appears to be an increasing interest in sufficiency and similar concepts, too, in particular in the area of transport planning, as not only the example of Copenhagen shows (cf. section 5.2.3). Hopefully, the renewed optimism and dynamic as a result of the outcome of the climate conference in Paris in December 2015, where at least ambitious targets were agreed on even if important details about how to get there are still lacking (cf. chapter 1; cf. UNFCCC 2015), can contribute to a development that allows for broader acceptance of policies that are in line with the idea of sufficiency, whether they use the label "sufficiency" or not.

References

Achterberg, W., 1993. Can liberal democracy survive the environmental crisis? Sustainability, liberal neutrality and overlapping consensus. In: A. Dobson and P. Lucardie, eds. The politics of nature: explorations in green political theory. London, New York: Routledge, 81–101.

Alcott, B., 2005. Jevons' paradox. Ecological Economics, 54 (1), 9–21.

Alcott, B., 2008. The sufficiency strategy: Would rich-world frugality lower environmental impact? Ecological Economics, 64 (4), 770–786.

Alexander, S., 2012. Living Better on Less? Toward an Economics of Sufficiency. Simplicity Institute, Simplicity Institute Report No. 12 c.

Anderson, E.S., 1999. What Is the Point of Equality? Ethics, 109 (2), 287–337.

Anderson, S., 2014. Coercion. The Stanford Encyclopedia of Philosophy (Spring 2014 Edition). http://plato.stanford.edu/archives/spr2014/entries/coercion/ (last accessed March 2014).

AnimalAid, no date. Eating Meat: Wrecking The Planet And Creating A Global Food Shortage. http://www.animalaid.org.uk/images/pdf/factfiles/wreck.pdf (last accessed October 2015).

Arneson, R., 1989. Equality and Equal Opportunity for Welfare. Philosophical Studies, 56, 77–93.

Arneson, R., 2006. Distributive Justice and Basic Capability Equality: 'Good Enough' is not Good Enough. In: A. Kaufman, ed. Capabilities Equality: Basic Issues and Problems. New York: Routledge, 17–43.

Arneson, R., 2013. Egalitarianism. The Stanford Encyclopedia of Philosophy (Summer 2013 Edition). http://plato.stanford.edu/archives/sum2013/entries/egalitarianism/ (last accessed September 2013).

Arneson, R.J., 2003. Liberal Neutrality on the Good: An Autopsy. In: S. Wall and G. Klosko, eds. Perfectionism and Neutrality: Essays in Liberal Theory. Lanham (etc.): Rowman and Littlefield, 191–218.

Arneson, R.J., 2004. Democracy is not intrinsically just. In: K. Dowding, R.E. Goodin, and C. Pateman, eds. Justice and Democracy: Essays for Brian Barry. Cambridge: Cambridge University Press, 40–58.

Arrow, K., Bolin, B., Costanza, R., Dasgupta, P., Folke, C., Holling, C.S., Jansson, B.-O., Levin, S., Mäler, K.-G., Perrings, C., and Pimentel, D., 1995. Economic Growth, Carrying Capacity, and the Environment. Science, 268, 520–521.

Arrow, K., Dasgupta, P., Goulder, L.H., Daily, G., Ehrlich, P.R., Heal, G., Levin, S., Mäler, K.-G., Schneider, S., Starrett, D., and Walker, B., 2004. Are we consuming too much? Journal of Economic Perspectives, 18 (3), 147–172.

Avnon, D. and de-Shalit, A., 1999. Introduction: Liberalism between promise and practice. In: D. Avnon and A. de-Shalit, eds. Liberalism and its Practice. London, New York: Routledge, 1–15.

Ayres, R.U. and Ayres, L.W., eds., 2002. A Handbook of Industrial Ecology. Cheltenham (etc.): Edward Elgar.

Ayres, R.U., 2007. On the practical limits to substitution. Ecological Economics, 61 (1), 115–128.

Baatz, C., 2014. Climate Change and Individual Responsibility to Reduce GHG Emissions. Ethics, Policy & Environment, 17 (1), 1–19.

Bailey, I., 2002. European Environmental Taxes and Charges: Economic Theory and Policy Practice. Applied Geography, 22, 235–251.

Balabanian, N., 1980. Presumed Neutrality of Technology. Society, 17 (3), 7–14.

Ball, T., 2006. Democracy. In: A. Dobson and R. Eckersley, eds. Political Theory and the Ecological Challenge. Cambridge, New York: Cambridge University Press, 131–147.

Barker, T., Dagoumas, A., and Rubin, J., 2009. The macroeconomic rebound effect and the world economy. Energy Efficiency, 2, 411–427.

Barry, B., 1996. Circumstances of Justice and Future Generations. In: R.I. Sikora and B. Barry, eds. Obligations to future generations. Cambridge: White Horse Press, 204–248.

Barry, B., 1999. Sustainability and Intergenerational Justice. In: A. Dobson, ed. Fairness and Futurity—Essays on Environmental Sustainability and Social Justice. Oxford, New York: Oxford University Press, 93–117.

Barry, J. and Doran, P., 2006. Refining Green Political Economy: From Ecological Modernisation to Economic Security and Sufficiency. Analyse & Kritik, 28, 250–275.

Barry, J., 1996. Sustainability, Political Judgement and Citizenship—Connecting green politics and democracy. In: B. Doherty and M. de Geus, eds. Democracy and Green Political Thought—Sustainability, rights and citizenship. London, New York: Routledge, 115–131.

Barry, J., 2001. Greening Liberal Democracy: Practice, Theory, and Political Economy. In: J. Barry and M. Wissenburg, eds. Sustaining liberal democracy. Ecological challenges and opportunities. New York (etc.): Palgrave, 59–80.

Bartelmus, P., 2002. Suffizienz und Nachhaltigkeit. In: M. Linz, P. Bartelmus, P. Hennicke, R. Jungkeit, W. Sachs, G. Scherhorn, G. Wilke, and U. v. Winterfeld, eds. Von nichts zu viel—Suffizienz gehört zur Zukunftsfähigkeit. 39–48.

Baumgärtner, S., Becker, C., Faber, M., and Manstetten, R., 2006. Relative and absolute scarcity of nature. Assessing the roles of economics and ecology for biodiversity conservation. Ecological Economics, 59, 487–498.

Baxter, B., 2006. Political Liberalism, the Non-Human Biotic and the Abiotic: A Response to Simon Hailwood. Analyse & Kritik, 28, 190–205.

Beckerman, W., 1999. Sustainable Development and our Obligations to Future Generations. In: A. Dobson, ed. Fairness and Futurity—Essays on Environmental Sustainability and Social Justice. Oxford, New York: Oxford University Press, 71–92.

Beemsterboer, S. and Kemp, R., 2016. Sustainability Assessment of Technologies. In: H. Heinrichs, P. Martens, G. Michelsen, and A. Wiek, eds. Sustainability Science: An Introduction. Heidelberg, New York, London: Springer, 71–83.

Bell, D.R., 2002. How can Political Liberals be Environmentalists? Political Studies, 50, 703–724.

Bell, D.R., 2006. Political Liberalism and Ecological Justice. Analyse & Kritik, 28, 206–222.

Berglund, C. and Matti, S., 2006. Citizen and consumer: the dual role of individuals in environmental policy. Environmental Politics, 15 (4), 550–571.

Berlin, I., 1969. Two Concepts of Liberty. In: Four Essays on Liberty. Oxford: Oxford University Press, 118–172.

Bilharz, M. and Fricke, V., 2011. Konsum und gesellschaftlicher Wandel: Das 'Großgedruckte' beachten. Zeitschrift für internationale Bildungsforschung und Entwicklungspädagogik, (4), 9–12.

Bilharz, M., Fricke, V., and Schrader, U., 2011. Wider die Bagatellisierung der Konsumentenverantwortung. GAIA, 20 (1), 9–13.

Binswanger, H.C., 2006. Die Wachstumsspirale. Marburg: Metropolis.

Binswanger, H.C., 2009. Vorwärts zur Mäßigung. Perspektiven einer nachhaltigen Wirtschaft. Hamburg: Murmann.

Birnbacher, D., 1977. Rawls' 'Theorie der Gerechtigkeit' und das Problem der Gerechtigkeit Zwischen den Generationen. Zeitschrift für Philosophische Forschung, 31 (3), 385–401.

Birnbacher, D., 1988. Verantwortung für zukünftige Generationen. Stuttgart: Reclam.

Blake, M. and Smith, P.T., 2013. International Distributive Justice. The Stanford Encyclopedia of Philosophy (Winter 2013 Edition). http://plato.stanford.edu/archives/win2013/entries/international-justice (last accessed November 2014).

BMU (Bundesministerium für Umwelt, Naturschutz und Reaktorsicherheit), 2008. Die Dritte industrielle Revolution—Aufbruch in ein ökologisches Jahrhundert. Dimensionen und Herausforderungen des industriellen und gesellschaftlichen Wandels. Berlin.

Boardman, B., 2010. Fixing Fuel Poverty: Challenges and Solutions. Oxon, New York: Earthscan.

Bohman, J. and Richardson, H.S., 2009. Liberalism, Deliberative Democracy, and "Reasons that All Can Accept". The Journal of Political Philosophy, 17 (3), 253–274.

Bookchin, M., 1989. Death of a small planet—It's growth that's killing us. The Progressive, 19–23.

Bosquet, B., 2000. Environmental tax reform: does it work? A survey of the empirical evidence. Environmental Economics, 34 (1), 19–32.

Boulanger, P.-M., 2010. Three strategies for sustainable consumption. S.A.P.I.EN.S., 3 (2).

Bracke, R., Albrecht, and De Clercq, 2006. The use of negotiated environmental agreements: from gentlemen's agreements to binding contracts. Ghent University, Working Paper No. 2006/415.

Brand, F., 2009. Critical natural capital revisited: Ecological resilience and sustainable development. Ecological Economics, 68 (3), 605–612.

Brand, K.-W. and Jochum, G., 2000. Der Deutsche Diskurs zu nachhaltiger Entwicklung. Abschlussbericht eines DFG-Projekts zum Thema „Sustainable Development/ Nachhaltige Entwicklung—Zur sozialen Konstruktion globaler Handlungskonzepte im Umweltdiskurs". München, MPS-Texte No. 1/2000.

Brand, U., 2012. Green Economy—the Next Oxymoron? No Lessons Learned from Failures of Implementing Sustainable Development. GAIA, 21 (1), 28–32.

Brink, D., 2008. Mill's Moral and Political Philosophy. The Stanford Encyclopedia of Philosophy (Fall 2008 Edition). http://plato.stanford.edu/archives/fall2008/entries/ mill-moral-political/ (last accessed August 2013).

Brischke, L.-A. and Spengler, L., 2011. Effizienz und Suffizienz—Ein Fall für zwei. Politische Ökologie, 126, 86–93.

Brischke, L.-A., Pehnt, M., and Jacobsen, S., 2011. Verankerung von Suffizienz im energiepolitischen Instrumentarium am Beispiel des Stromverbrauchs. Unpublished Paper, Heidelberg.

Bristow, A.L., Wardman, M., Zanni, A.M., and Chintakayala, P.K., 2010. Public acceptability of personal carbon trading and carbon tax. Ecological Economics, 69, 1824–1837.

Brook, B.W. and Bradshaw, C.J.A., 2012. Strange bedfellows? Techno-fixes to solve the big conservation issues in southern Asia. Biological Conservation, 151, 7–10.

Brown, P.M. and Cameron, L.D., 2000. What can be done to reduce overconsumption? Ecological Economics, 32, 27–41.

BUND (Bund für Umwelt und Naturschutz Deutschland), Brot für die Welt, and Evangelischer Entwicklungsdienst, eds., 2008. Zukunftsfähiges Deutschland in einer globalisierten Welt. Bonn: Bundeszentrale für politische Bildung.

Bundesregierung, 2002. Perspectives for Germany. Our Strategy for Sustainable Development. Berlin.

Bundesregierung, 2009. Nationaler Entwicklungsplan Elektromobilität. http://www.bmvi.de/SharedDocs/DE/Anlage/VerkehrUndMobilitaet/nationaler-entwicklungsplan-elektromobilitaet.pdf?__blob=publicationFile (last accessed November 2015).

Bundesregierung, 2016. Nationales Programm für nachhaltigen Konsum. http://www.bmub.bund.de/fileadmin/Daten_BMU/Download_PDF/Produkte_und_Umwelt/nat_programm_konsum_bf.pdf (last accessed March 2016).

Bündnis 90/Die Grünen, 2013. Zeit für den grünen Wandel. Teilhaben. Einmischen. Zukunft schaffen. Bundestagswahlprogramm 2013. Berlin.

Byron, M., 2004. Introduction. In: M. Byron, ed. Satisficing and Maximizing: Moral Theorists on Practical Reason. Cambridge (etc.): Cambridge University Press, 1–13.

Cairns, S., Sloman, L., Newson, C., Anable, J., Kirkbride, A., and Goodwin, P., 2008. Smarter Choices: Assessing the Potential to Achieve Traffic Reduction Using 'Soft Measures'. Transport Reviews, 28 (5), 593–618.

Calder, G. and McKinnon, C., 2011. Introduction: Climate change and liberal priorities. Critical Review of International Social and Political Philosophy, 14 (2), 91–97.

Campbell, C., 1994. Consuming goods and the good of consuming. Critical Review, 8 (4), 503–520.

Caney, S., 2003. Entitlements, Obligations, and Distributive Justice—The Global Level. In: D.A. Bell and A. de-Shalit, eds. Forms of Justice: Critical Perspectives on David Miller's Political Philosophy. Lanham: Rowman and Littlefield, 287–313.

Caney, S., 2005. Cosmopolitan justice, responsibility, and global climate change. Leiden Journal of International Law, 18 (4), 747–775.

Caney, S., 2009. Justice and the distribution of greenhouse gas emissions. Journal of Global Ethics, 5 (2), 125–146.

Caney, S., 2010. Climate Change, Human Rights, and Moral Thresholds. In: S.M. Gardiner, S. Caney, D. Jamieson, and H. Shue, eds. Climate Ethics—Essential Readings. Oxford (etc.): Oxford University Press, 163–172.

Carter, I., 2012. Positive and Negative Liberty. The Stanford Encyclopedia of Philosophy (Spring 2012 Edition). http://plato.stanford.edu/archives/spr2012/entries/liberty-positive-negative/ (last accessed August 2013).

Casal, P., 2007. Why sufficiency is not enough. Ethics, 117 (2), 296–326.

Christen, M. and Schmidt, S., 2011. A Formal Framework for Conceptions of Sustainability—A Theoretical Contribution to the Discourse in Sustainable Development. Sustainable Development (online edition).

Christiano, T., 2008. Democracy. The Stanford Encyclopedia of Philosophy (Fall 2008 Edition). http://plato.stanford.edu/archives/fall2008/entries/democracy/ (last accessed June 2013).

Christiano, T., ed., 2003. Philosophy and Democracy: An Anthology. Oxford (etc.): Oxford University Press.

Christman, J. and Anderson, J., eds., 2005. Autonomy and the Challenges to Liberalism—New Essays. Cambridge: Cambridge University Press.

Ciriacy-Wantrup, S. v., 1952. Resource Conservation: Economics and Policies. Berkeley, Los Angeles: University of California Press.

City of Copenhagen, 2015. Copenhagen—City of Cyclists. The Bicycle Account 2014. Copenhagen.

Clark, W.C., 2007. Sustainability Science: A room of its own. Proceedings of the National Academy of Sciences, 104 (6), 1737–1738.

Clark, W.C., Crutzen, P.J., and Schellnhuber, H.J., 2004. Science for Global Sustainability—Toward a New Paradigm. In: H.J. Schellnhuber, P.J. Crutzen, and W.C. Clark, eds. Earth System Analysis for Sustainability. Cambridge (Mass.): MIT Press, 1–28.

Coglianese, C., 1998. Implications of Liberal Neutrality for Environmental Policy. Environmental Ethics, 20 (1), 41–59.

Cohen, G.A., 1989. On the Currency of Egalitarian Justice. Ethics, 99 (4), 906–944.

Cohen, G.A., 2008. Rescuing Justice and Equality. Cambridge (etc.): Harvard University Press.

Cohen, G.A., 2009. Why not socialism? Princeton, New Jersey (etc.): Princeton University Press.

Cohen, J. and Sabel, C., 2006. Extra Republicam Nulla Justitia? Philosophy & Public Affairs, 34 (2), 147–175.

Cohen, J., 2003. Procedure and Substance in Deliberative Democracy. In: T. Christiano, ed. Philosophy and Democracy: An Anthology. Oxford (etc.): Oxford University Press, 17–38.

Cohen, M.J., 2011. Is the UK preparing for 'war'? Military metaphors, personal carbon allowances, and consumption rationing in historical perspective. Climatic Change, 104 (2), 199–222.

Costanza, R., Alperovitz, G., Daly, H.E., Farley, J., Franco, C., Jackson, T., Kubiszewski, I., Schor, J., and Victor, P., 2012. Building a Sustainable and Desirable Economy-in-Society-in-Nature. New York: United Nations Division for Sustainable Development.

Costanza, R., d'Arge, R., de Groot, R., Farber, S., Grasso, M., Hannon, B., Limburg, K., Naeem, S., O'Neill, R.V., Paruelo, J., Raskin, R.G., Sutton, P., and van den Belt, M., 1997. The value of the world's ecosystem services and natural capital. Nature, 387, 253–260.

Council of the European Union, 2006. Review of the EU Sustainable Development Strategy (EU SDS)—Renewed Strategy, 10917/06.

Cranor, C.F., 1985. Collective and individual duties to protect the environment. Journal of Applied Philosophy, 2 (2), 243–259.

Cripps, E., 2011. Climate change, collective harm and legitimate coercion. Critical Review of International Social and Political Philosophy, 14 (2), 171–193.

Crisp, R., 2003. Equality, Priority, and Compassion. Ethics, 113 (4), 745–763.

Croson, R. and Treich, N., 2014. Behavioral Environmental Economics: Promises and Challenges. Environmental and Resource Economics, 58 (3), 335–351.

Dahl, J., 1991. Zwölfzylinder, schadstoffarm. Von den Aporien des sogenannten Umweltschutzes. Scheidewege, 21, 228–245.

Dahmus, J.B., 2014. Can Efficiency Improvements Reduce Resource Consumption? A Historical Analysis of Ten Activities. Journal of Industrial Ecology, 18 (6), 883–897.

Daly, H.E., 1977. Steady-State Economics. The Economics of Biophysical Equilibrium and Moral Growth. San Francisco: W. H. Freeman and Company.

Daly, H.E., 1991. Steady-state economics. 2nd edn, with new essays. Washington, DC (etc.): Island Press.

Daly, H.E., 1993. The Steady-State Economy: Toward a Political Economy of Biophysical Equilibrium and Moral Growth. In: H.E. Daly and K.N. Townsend, eds. Valuing the earth: economics, ecology, ethics. Cambridge, MA (etc.): MIT Press, 325–363.

Daly, H.E., 1996. Beyond Growth: the economics of sustainable development. Boston (Mass.): Beacon Press.

Daly, H.E., Czech, B., Trauger, D.L., Rees, W.E., Grover, M., Dobson, T., and Trombulak, S.C., 2007. Are We Consuming Too Much—for What? Conservation Biology, 21 (5), 1359–1362.

Daniels, N., 1990. Equality of What: Welfare, Resources, or Capabilities? Philosophy and Phenomenological Research, 50, 273–296.

Darby, S., 2007. Enough is as good as a feast—sufficiency as policy. In: S. Attali and K. Tillerson, eds. Saving Energy—Just do it! ECEEE 2007 Summer Study. Stockholm, 111–119.

Dauvergne, P., 2010. The Problem of Consumption. Global Environmental Politics, 10 (2), 1–10.

de Geus, M., 2001. Sustainability, Liberal Democracy, Liberalism. In: J. Barry and M. Wissenburg, eds. Sustaining liberal democracy. Ecological challenges and opportunities. New York (etc.): Palgrave, 19–36.

de Geus, M., 2004. The environment versus individual freedom and convenience. In: M. Wissenburg and Y. Levy, eds. Liberal democracy and environmentalism: The end of environmentalism? London, New York: Routledge, 87–99.

De Groot, R., Van der Perk, J., Chiesura, A., and van Vliet, A., 2003. Importance and threat as determining factors for criticality of natural capital. Ecological Economics, 44 (2–3), 187–204.

de-Shalit, A., 1995. Is Liberalism Environment-Friendly? Social Theory and Practice, 21 (2), 287–314.

de-Shalit, A., 1997. Environmentalism for Europe—One Model? Journal of Applied Philosophy, 14 (2), 177–186.

de-Shalit, A., 2000. The Environment: Between Theory and Practice. Oxford, New York: Oxford University Press.

Defra (Department for Environment, Food and Rural Affairs), 2008. An assessment of the potential effectiveness and strategic fit of personal carbon trading. London: Defra.

Delmas, M.A., Fischlein, M., and Asensio, O.I., 2013. Information strategies and energy conservation behavior: A meta-analysis of experimental studies from 1975 to 2012. Energy Policy, 61, 729–739.

Deutscher Bundestag, 2013. Schlussbericht der Enquete-Kommission „Wachstum, Wohlstand, Lebensqualität—Wege zu nachhaltigem Wirtschaften und gesellschaftlichem Fortschritt in der Sozialen Marktwirtschaft". Bonn: Bundeszentrale für politische Bildung.

Diefenbacher, H. and Habicht-Erenler, S., eds., 1991. Wachstum und Wohlstand: neuere Konzepte zur Erfassung der Sozial- und Umweltverträglichkeit. Marburg: Metropolis.

Diefenbacher, H., 2007. Wirtschaftswachstum als Statistik-Phantom. In: S. Rudolph, ed. Wachstum, Wachstum über alles? Ein ökonomisches Leitbild auf dem Prüfstand von Umwelt und Gerechtigkeit. Marburg: Metropolis, 31–51.

Diefenbacher, H., 2008. Eine Dritte industrielle Revolution—Die große schöne Lösung für das 21. Jahrhundert? In: BMU (Bundesministerium für Umwelt, Naturschutz und Reaktorsicherheit), ed. Die Dritte industrielle Revolution—Aufbruch in ein ökologisches Jahrhundert. Dimensionen und Herausforderungen des industriellen und gesellschaftlichen Wandels. Berlin, 130–132.

Diefenbacher, H., Zieschank, R., and Rodenhäuser, D., 2010. Measuring Welfare in Germany: A suggestion for a new welfare index. Dessau-Roßlau: Umweltbundesamt, UBA Texte No. 03/2010.

Diekmann, J., 1999. Zwischen Effizienz und Suffizienz. Ökologisches Wirtschaften, 3/1999, 25–26.

DiLorenzo, T., 2011. The Trojan Horse of 'Happiness Research'. https://mises.org/library/trojan-horse-happiness-research (last accessed May 2015).

Dinda, S., 2004. Environmental Kuznets Curve Hypothesis: A Survey. Ecological Economics, 49, 431–455.

Dobson, A. and Lucardie, P., eds., 1993. The politics of nature: explorations in green political theory. London, New York: Routledge.

Dobson, A., 1996 a. Democratising Green Theory—Preconditions and principles. In: B. Doherty and M. de Geus, eds. Democracy and Green Political Thought—Sustainability, rights and citizenship. London, New York: Routledge, 132–148.

Dobson, A., 1996 b. Environmental Sustainabilities: An Analysis and a Typology. Environmental Politics, 5 (3), 401–428.

Dobson, A., 2007. Green Political Thought. 4. ed. London, New York: Routledge.

Dobson, A., 2010. Environmental citizenship and pro-environmental behaviour. London: Sustainable Development Research Network.

Doherty, B. and de Geus, M., eds., 1996. Democracy and Green Political Thought—Sustainability, rights and citizenship. London, New York: Routledge.

Donaldson, S. and Kymlicka, W., 2011. Zoopolis: A Political Theory of Animal Rights. Oxford: Oxford University Press.

Dowding, K., Goodin, R.E., and Pateman, C., eds., 2004. Justice and Democracy: Essays for Brian Barry. Cambridge: Cambridge University Press.

Doyal, L. and Gough, I., 1991. A Theory of Human Need. Basingstoke, Hampshire (etc.): Macmillan.

Dryzek, J.S. and Dunleavy, P., 2009. Theories of the Democratic State. Basingstoke, Hampshire (etc.): Palgrave Macmillan.

Dryzek, J.S., 1992. Ecology and discursive democracy: Beyond liberal capitalism and the administrative state. Capitalism, Nature, Socialism, 3 (2), 18–42.

Dryzek, J.S., 2005. The Politics of the Earth: Environmental Discourses. 2nd edn Oxford, New York: Oxford University Press.

Durning, A.T., 1992. How much is enough? The consumer society and the future of the Earth. London: Earthscan.

Dworkin, G., 2010. Paternalism. The Stanford Encyclopedia of Philosophy (Summer 2010 Edition). http://plato.stanford.edu/archives/sum2010/entries/paternalism/ (last accessed August 2013).

Dworkin, R., 1978. Liberalism. In: S. Hampshire, ed. Public and Private Morality. Cambridge (etc.): Cambridge University Press, 113–143.

Dworkin, R., 1981 a. What is Equality? Part 1: Equality of Welfare. Philosophy & Public Affairs, 10 (3), 185–246.

Dworkin, R., 1981 b. What is Equality? Part 2: Equality of Resources. Philosophy & Public Affairs, 10 (4), 283–345.

Dworkin, R., 1987. What is Equality? Part 3: The Place of Liberty. Iowa Law Review, 73 (1), 1–54.

Dworkin, R., 2011. Justice for Hedgehogs. Cambridge (Mass.) (etc.): Belknap Press of Harvard University Press.

Easterlin, R.A., 2001. Income and Happiness: Towards a Unified Theory. The Economic Journal, 111 (473), 465–484.

EC (European Commission), 2000. Communication from the Commission on the precautionary principle, COM(2000) 1 final. Brussels.

EC (European Commission), 2001. Communication from the Commission: A Sustainable Europe for a Better World: A European Union Strategy for Sustainable Development, COM(2001)264 final. Brussels.

EC (European Commission), 2005. Communication from the Commission to the Council, the European Parliament, the European Economic and Social Committee and the Committee of the Regions: Thematic Strategy on the sustainable use of natural resources, COM(2005) 670 final. Brussels.

EC (European Commission), 2008. Communication from the Commission to the European Parliament, the Council, the European Economic and Social Committee and the Committee of the Regions on the Sustainable Consumption and Production and Sustainable Industrial Policy Action Plan, COM(2008) 397 final. Brussels.

EC (European Commission), 2010. Communication from the Commission—EUROPE 2020: A strategy for smart, sustainable and inclusive growth, COM(2010) 2020 final. Brussels.

EC (European Commission), 2015. Communication from the Commission to the European Parliament, the Council, the European Economic and Social Committee and the Committee of the Regions: Closing the loop—An EU action plan for the Circular Economy, COM(2015) 614 final. Brussels.

EC (European Commission), 2016. Communication from the Commission to the European Parliament, the Council, the European Economic and Social Committee and the Committee of the Regions: A European agenda for the collaborative economy, COM(2016) 356 final. Brussels.

Eckersley, R., 1996. Greening Liberal Democracy—The rights discourse revisited. In: B. Doherty and M. de Geus, eds. Democracy and Green Political Thought—Sustainability, rights and citizenship. London, New York: Routledge, 212–236.

EEA (European Environment Agency), 2010. The European Environment—State and Outlook 2010: Consumption and the Environment. Copenhagen.

EEA (European Environment Agency), 2012 a. Environmental Indicator Report 2012 —Ecosystem Resilience and Resource Efficiency in a Green Economy in Europe. Copenhagen.

EEA (European Environment Agency), 2012 b. The European Environment State and Outlook 2010—Consumption and Environment 2012 Update. Copenhagen.

EEA (European Environment Agency), 2013 a. Environmental pressures from European consumption and production—A study in integrated environmental and economic analysis. EEA Technical report No. 2/2013, Copenhagen.

EEA (European Environment Agency), 2013 b. A closer look at urban transport. TERM 2013: transport indicators tracking progress towards environmental targets in Europe. EEA Report No. 11/2013, Copenhagen.

EEA (European Environment Agency), 2014. Environmental Indicator Report 2014—Environmental Impacts of Production-Consumption Systems in Europe. Copenhagen.

EEA (European Environment Agency), 2015. The European Environment—State and Outlook 2015: Synthesis Report. Copenhagen.

Ehrlich, P.R. and Goulder, L.H., 2007. Is Current Consumption Excessive? A General Framework and Some Indications for the United States. Conservation Biology, 21 (5), 1145–1154.

Ehrlich, P.R. and Holdren, J.P., 1971. Impact of Population Growth. Science, 171 (3977), 1212–1217.

Ekardt, F., 2016. Suffizienz: Politikinstrumente, Grenzen von Technik und Wachstum und die schwierige Rolle des guten Lebens. Soziologie und Nachhaltigkeit, (04/2016), 1–25.

Ekins, P. and Speck, S., 2000. Proposals of environmental fiscal reforms and the obstacles to their implementation. Journal of Environmental Policy & Planning, 2 (2), 93–114.

Ekins, P., 1993. 'Limits to growth' and 'sustainable development': grappling with ecological realities. Ecological Economics, 8, 269–288.

Ekins, P., 1999. European Environmental Taxes and Charges: Recent Experience, Issues and Trends. Ecological Economics, 31, 39–62.

Ekins, P., 2003. Identifying critical natural capital—Conclusions about critical natural capital. Ecological Economics, 44 (2–3), 277–292.

Ekins, P., Simon, S., Deutsch, L., Folke, C., and De Groot, R., 2003. A framework for the practical application of the concepts of critical natural capital and strong sustainability. Ecological Economics, 44 (2–3), 165–185.

Elgin, D. and Mitchell, A., 1977. Voluntary Simplicity. The Co-Evolution Quarterly, Summer 1977. http://www.duaneelgin.com/wp-content/uploads/2010/11/voluntary_simplicity.pdf (last accessed October 2012).

Elkington, J., 1997. Cannibals with forks: the triple bottom line of 21st century business. Chichester: Capstone Pub.

Elster, J., 1986. The market and the forum: three varieties of political theory. In: J. Elster and A. Hylland, eds. Foundations of Social Choice Theory. Cambridge (etc.): Cambridge University Press, 103–132.

Elster, J., 1998. Introduction. In: J. Elster, ed. Deliberative Democracy. Cambridge (etc.): Cambridge University Press, 1–18.

Enquete-Kommission, 1994. Die Industriegesellschaft gestalten—Perspektiven für einen nachhaltigen Umgang mit Stoff- und Materialströmen. Bericht der Enquete-Kommission 'Schutz des Menschen und der Umwelt—Bewertungskriterien und Perspektiven für umweltverträgliche Stoffkreisläufe in der Industriegesellschaft'. Bonn.

Enquete-Kommission, 1998. Konzept Nachhaltigkeit: vom Leitbild zur Umsetzung. Abschlussbericht der Enquete-Kommission 'Schutz des Menschen und der Umwelt —Ziele und Rahmenbedingungen einer nachhaltig zukunftsverträglichen Entwicklung'. Bonn.

EPA Network (Network of Heads of European Environment Protection Agencies), 2006. Delivering the sustainable use of natural resources: A contribution from the following members of the Network of Heads of European Environment Protection Agencies on the Thematic Strategy on the Sustainable Use of Natural Resources.

Eshel, G., Shepon, A., Makov, T., and Milo, R., 2014. Land, irrigation water, greenhouse gas, and reactive nitrogen burdens of meat, eggs, and dairy production in the United States. Proceedings of the National Academy of Sciences, PNAS Early Edition.

Eshleman, A., 2009. Moral Responsibility. The Stanford Encyclopedia of Philosophy (Winter 2009 Edition). http://plato.stanford.edu/archives/win2009/entries/moral-responsibility/ (last accessed January 2013).

FAO (Food and Agriculture Organization of the United Nations), 2014. The State of the World Fisheries and Aquaculture—Opportunities and Challenges. Rome: FAO Fisheries and Aquaculture Department.

Fawcett, T., 2010. Personal carbon trading: A policy ahead of its time? Energy Policy, 38, 6868–6876.

Fawcett, T., Bottrill, C., Boardman, B., and Lye, G., 2007. Trialling personal carbon allowances. London: UK Energy Research Centre.

Feinberg, J., 1984. Harm to Others—The Moral Limits of the Criminal Law. New York, Oxford: Oxford University Press.

Felder, S. and Schleiniger, R., 2002. Environmental Tax Reform: Efficiency and Political Feasibility. Ecological Economics, 42, 107–116.

Fischer, C. and Grießhammer, R., 2013. Mehr als nur weniger—Suffizienz: Begriff, Begründung und Potenziale. Öko-Institut, Öko-Institut Working Paper No. 2/2013.

Fishman, E., Washington, S., and Haworth, N., 2014. Bike share's impact on car use: Evidence from the United States, Great Britain, and Australia. Transportation Research Part D, 31, 13–20.

Focus Online, 2013. Arrogante „Verbotsmentalität" ärgert selbst Ur-Grüne, 5 Aug. http://www.focus.de/politik/deutschland/tid-32754/die-lizenz-zum-vorschreiben-veggie-tag-tueten-abgabe-tempolimit-die-bevormundungs-wut-der-gruenen_aid_1063304.html (last accessed August 2013).

FoEE (Friends of the Earth Europe), ed., 1995. Towards Sustainable Europe—The Study. Brussels.

Folke, C., Carpenter, S., Walker, B., Scheffer, M., Elmqvist, T., Gunderson, L., and Holling, C.S., 2004. Regime shifts, resilience and biodiversity in ecosystem management. Annual Review in Ecology, Evolution and Systematics, 35, 557–581.

Fraenkel, E., 1973. Deutschland und die westlichen Demokratien. Stuttgart (etc.): Verlag W. Kohlhammer.

Frankfurt, H., 1987. Equality as a Moral Ideal. Ethics, 98 (1), 21–43.

Frankfurt, H., 1997. Equality and Respect. Social Research, 64 (1), 3–15.

Frankfurt, H.G., 1984. Necessity and Desire. Philosophy and Phenomenological Research, 45 (1), 1–13.

Freeden, M., 2008. European Liberalisms: An Essay in Comparative Political Thought. European Journal of Political Theory, 7 (1), 9–30.

Freedom House, 2015 a. Freedom in the World 2015—Discarding Democracy: Return to the Iron Fist. Washington, New York: Freedom House.

Freedom House, 2015 b. Freedom in the World 2015 Methodology. Washington, New York: Freedom House.

Frevel, B., 2004. Demokratie: Entwicklung, Gestaltung, Problematisierung. Wiesbaden: VS Verlag für Sozialwissenschaften.

Frey, B.S. and Stutzer, A., 2002. Happiness and Economics. Princeton, Oxford: Princeton University Press.

Fromm, E., 1976. Haben oder Sein: Die seelischen Grundlagen einer neuen Gesellschaft. Stuttgart: Deutsche Verlags-Anstalt.

Fudge, S. and Peters, M., 2011. Behaviour Change in the UK Climate Debate: An Assessment of Responsibility, Agency and Political Dimensions. Sustainability, 3, 789–808.

Galbraith, J.K., 1958 a. The Affluent Society. London: Hamilton.

Galbraith, J.K., 1958 b. How Much Should a Country Consume? In: H. Jarrett, ed. Perspectives on Conservation—Essays on America's Natural Resources. Baltimore, Maryland: Johns Hopkins Press, 89–99.

Galston, W.A., 2007. Why the new liberalism isn't all that new, and why the old liberalism isn't what we thought it was. Social Philosophy and Policy, 24 (1), 289–305.

Gärling, T. and Schuitema, G., 2007. Travel Demand Management Targeting Reduced Private Car Use: Effectiveness, Public Acceptability and Political Feasibility. Journal of Social Issues, 63 (1), 139–153.

Gaus, G. and Courtland, S.D., 2010. Liberalism. The Stanford Encyclopedia of Philosophy. http://plato.stanford.edu/entries/liberalism/ (last accessed April 2013).

Gaus, G., 2005. The Place of Autonomy within Liberalism. In: J. Christman and J. Anderson, eds. Autonomy and the Challenges to Liberalism—New Essays. Cambridge: Cambridge University Press, 272–306.

Georgescu-Roegen, N., 1975. Energy and Economic Myths. Southern Economic Journal, 41 (3), 347–381.

Getzner, M., 2009. Wirtschaftswachstum und Umweltverbrauch: Über die Zusammenhänge von Energie- und Ressourcenverbrauch, Emissionen und BIP in Österreich. Wissenschaft & Umwelt Interdisziplinär, 13, 22–31.

Giddens, A., 1984. The Constitution of Society: Outline of the Theory of Structuration. Cambridge (UK): Polity Press.

Gillingham, K., Kotchen, M.J., Rapson, D.S., and Wagner, G., 2013. Energy policy: The rebound effect is overplayed. Nature, 493 (7433), 475–476.

Golub, A., 2016. Mobility and Sustainability. In: H. Heinrichs, P. Martens, G. Michelsen, and A. Wiek, eds. Sustainability Science: An Introduction. Heidelberg, New York, London: Springer, 261–272.

Goodin, R.E. and Reeve, A., eds., 1989. Liberal Neutrality. London, New York: Routledge.

Goodin, R.E., 1988. Reasons for Welfare: The Political Theory of the Welfare State. Princeton, New Jersey: Princeton University Press.

Goodin, R.E., 1990. Relative Needs. In: A. Ware and R.E. Goodin, eds. Needs and Welfare. London (etc.): Sage Publications, 12–33.

Goodin, R.E., 1992. Green Political Theory. Cambridge, Oxford: Polity Press.

Goodin, R.E., 1995. Utilitarianism as a Public Philosophy. Cambridge (etc.): Cambridge University Press.

Goodland, R. and Daly, H.E., 1996. Environmental Sustainability: Universal and Non-Negotiable. Ecological Applications, Vol. 6 (No. 4), 1002–1017.

Gosepath, S., 2003. Verteidigung egalitärer Gerechtigkeit. Deutsche Zeitschrift für Philosophie, 51 (2), 257–297.

Gosepath, S., 2011. Equality. The Stanford Encyclopedia of Philosophy (Spring 2011 Edition). http://plato.stanford.edu/archives/spr2011/entries/equality/ (last accessed August 2013).

Gössling, S. and Choi, A.S., 2015. Transport transitions in Copenhagen: Comparing the cost of cars and bicycles. Ecological Economics, 113, 106–113.

Gray, J., 1996. Mill on Liberty: A Defence. Second Edition. London, New York: Routledge.

Gray, J., 2004. Back to Mill: Review of 'The snake that swallowed its tail: some contradictions in modern liberalism' by Mark Garnett. The New Statesman, 15 Nov.

Gregg, R.B., 1936. The Value Of Voluntary Simplicity. Wallingford, Pennsylvania: Pendle Hill.

Grober, U., 2002. Tiefe Wurzeln: Eine kleine Begriffsgeschichte von 'Sustainable Development'—Nachhaltigkeit. Natur und Kultur, 3 (1), 116–128.

Grober, U., 2010. Die Entdeckung der Nachhaltigkeit: Kulturgeschichte eines Begriffs. München: Kunstmann.

Grunwald, A., 2004. Nachhaltigkeit begreifen: Zwischen Leitbild und Trugbild. GAIA, 13 (1).

Grunwald, A., 2006. Technikfolgenabschätzung als Nachhaltigkeitsbewertung—Konzeptionelle Herausforderungen und methodische Probleme. In: J. Kopfmüller, ed. Ein Konzept auf dem Prüfstand—Das integrative Nachhaltigkeitskonzept in der Forschungspraxis. Berlin: Edition Sigma, 39–61.

Grunwald, A., 2008. Technik und Politikberatung. Frankfurt am Main: Suhrkamp.

Grunwald, A., 2010. Wider die Privatisierung der Nachhaltigkeit. Warum ökologisch korrekter Konsum die Umwelt nicht retten kann. GAIA, 19 (3), 178–182.

Grunwald, A., 2011. Statt Privatisierung: Politisierung der Nachhaltigkeit. GAIA, 20 (1), 17–19.

Gutmann, A. and Thompson, D., 2004. Why deliberative democracy? Princeton, New Jersey (etc.): Princeton University Press.

Habermas, J., 1996. Drei normative Modelle der Demokratie. In: Die Einbeziehung des Anderen. Studien zur politischen Theorie. Frankfurt a. M.: Suhrkamp.

Hailwood, S., 2004. How to be a Green Liberal—Nature, Value and Liberal Philosophy. Chesham: Acumen.

Hamilton, L.A., 2003. The Political Philosophy of Needs. Cambridge (etc.): Cambridge University Press.

Hammitt, J.K., 2000. Are The Costs of Proposed Environmental Regulations Overestimated? Evidence from the CFC Phaseout. Environmental and Resource Economics, 16 (3), 281–301.

Hannis, M., 2005. Public Provision of Environmental Goods: Neutrality or Sustainability? A Reply to David Miller. Environmental Politics, 14 (5), 577–595.

Hardin, G., 1968. The Tragedy of the Commons. Science, 162, 1243–1248.

Hardin, R., 1993. Liberalism: Political and Economic. Social Philosophy and Policy, 10 (2), 121–144.

Hawken, P., Lovins, A.B., and Lovins, L.H., 1999. Natural Capitalism: Creating the Next Industrial Revolution. Boston (Mass.): Little, Brown and Co.

Hawkes, J., 2001. The Fourth Pillar of Sustainability—Culture's essential role in public planning. Cultural Development Network Victoria, Australia.

Hayek, F.A., 1960. The Constitution of Liberty. London: Routledge & Kegan Paul.

Hayward, T., 1994. Ecological Thought. An Introduction. Cambridge (UK): Polity Press.

Hayward, T., 1998. Political Theory and Ecological Values. New York: St. Martin's Press.

Hayward, T., 2007. Human Rights Versus Emissions Rights: Climate Justice and the Equitable Distribution of Ecological Space. Ethics & International Affairs, 21 (4), 431–450.

Heath, J., 2005. Liberal Autonomy and Consumer Sovereignty. In: J. Christman and J. Anderson, eds. Autonomy and the Challenges to Liberalism—New Essays. Cambridge: Cambridge University Press, 204–225.

Heidbrink, L. and Reidel, J., 2011. Nachhaltiger Konsum durch politische Selbstbindung. GAIA, 20 (3), 152–156.

Heidbrink, L. and Schmidt, I., 2011. Das Prinzip der Konsumentenverantwortung—Grundlagen, Bedingungen und Umsetzungen verantwortlichen Konsums. In: L. Heidbrink, I. Schmidt, and B. Ahaus, eds. Die Verantwortung des Konsumenten. Über das Verhältnis von Markt, Moral und Konsum. Frankfurt, New York: Campus Verlag, 25–56.

Heidbrink, L., Schmidt, I., and Ahaus, B., eds., 2011. Die Verantwortung des Konsumenten. Über das Verhältnis von Markt, Moral und Konsum. Frankfurt, New York: Campus Verlag.

Heinrich-Böll-Stiftung, ed., 2009. Auf dem Weg zu einem Green New Deal—Die Klima- und die Wirtschaftskrise als transatlantische Herausforderungen.

Heinrichs, H., 2013 a. Nachhaltigkeitspolitik: Neuer Kontext für Entscheidungen unter Unsicherheit und Risiko. In: R. von Detten, F. Faber, and M. Bemmann, eds. Unberechenbare Umwelt—Zum Umgang mit Unsicherheit und Nicht-Wissen. Wiesbaden: Springer, 219–252.

Heinrichs, H., 2013 b. Sharing Economy: A Potential New Pathway to Sustainability. GAIA, 22 (4), 228–231.

Heinrichs, H., Martens, P., Michelsen, G., and Wiek, A., eds., 2016. Sustainability Science: An Introduction. Heidelberg, New York, London: Springer.

Hellmann, K.-U., 2011. Der Eigensinn der Konsumenten—Zur Frage der Verantwortung bei der Zurechnung von Konsumentenverantwortung. In: L. Heidbrink, I. Schmidt, and B. Ahaus, eds. Die Verantwortung des Konsumenten. Über das Verhältnis von Markt, Moral und Konsum. Frankfurt, New York: Campus Verlag, 269–283.

Hennicke, P., 2002. Effizienz und Suffizienz in einem System nachhaltiger Energienutzung. In: M. Linz, P. Bartelmus, P. Hennicke, R. Jungkeit, W. Sachs, G. Scherhorn, G. Wilke, and U. v. Winterfeld, eds. Von nichts zu viel—Suffizienz gehört zur Zukunftsfähigkeit. 57–70.

Hertwich, E.G., 2005. Consumption and Industrial Ecology. Journal of Industrial Ecology, 9 (1-2), 1–6.

Heyen, D.A., Fischer, C., Barth, R., Brunn, C., Grießhammer, R., Keimeyer, F., and Wolff, F., 2013. Mehr als nur weniger—Suffizienz: Notwendigkeit und Optionen politischer Gestaltung. Öko-Institut, Working Paper No. 3/2013.

Hoeschele, W., 2007. Scarcity. Encyclopedia of Environment and Society.

Holzinger, H., 2010. Wirtschaften jenseits von Wachstum? Befunde und Ausblicke. Wien: Österreichisches Lebensministerium.

Höpflinger, F., 2010. Alterssicherungssysteme: Doppelte Herausforderung von demografischer Alterung und Postwachstum. In: I. Seidl and A. Zahrnt, eds. Postwachstumsgesellschaft. Konzepte für die Zukunft. Marburg: Metropolis, 53–63.

Hopwood, B., Mellor, M., and O'Brien, G., 2005. Sustainable Development: Mapping Different Approaches. Sustainable Development, 13 (1), 38–52.

Horstkötter, D., 2004. Sustainability and plurality: From the moderate end of the liberal equilibrium to the open end of a situated liberal neutrality. In: M. Wissenburg and Y. Levy, eds. Liberal democracy and environmentalism: The end of environmentalism? London, New York: Routledge, 155–166.

Hösle, V., 1991. Philosophie der ökologischen Krise. München: Beck.

Howaldt, J. and Schwarz, M., 2010. Soziale Innovation im Fokus: Skizze eines gesellschaftstheoretisch inspirierten Forschungskonzepts. Bielefeld: transcript Verlag.

Huber, J., 1995 a. Nachhaltige Entwicklung. Strategien für eine ökologische und soziale Erdpolitik. Berlin: Edition Sigma.

Huber, J., 1995 b. Nachhaltige Entwicklung durch Suffizienz, Effizienz und Konsistenz. In: P. Fritz, J. Huber, and H.W. Levi, eds. Nachhaltigkeit in naturwissenschaftlicher und sozialwissenschaftlicher Perspektive. Stuttgart: Hirzel, 31–46.

Huber, J., 1999. Industrielle Ökologie. Konsistenz, Effizienz und Suffizienz in zyklusanalytischer Betrachtung.

Huber, J., 2000. Towards industrial ecology: sustainable development as a concept of ecological modernization. Journal of Environmental Policy & Planning, 2 (4), 269–285.

Huber, J., 2008. Technological Environmental Innovations (TEIs) in a Chain-Analytical and Life-Cycle-Analytical Perspective. Journal of Cleaner Production, 16 (18), 1980–1986.

Huber, J., 2011. Ökologische Modernisierung und Umweltinnovation. In: M. Groß, ed. Handbuch Umweltsoziologie. Wiesbaden: VS Verlag für Sozialwissenschaften, 279–302.

Hubin, D.C., 1989. Scarcity and the Demands of Justice. Capital University Law Review, 18, 185–199.

Hueting, R. and Reijnders, L., 1998. Sustainability is an objective concept. Ecological Economics, 27, 139–147.

Hueting, R., 1980. New Scarcity and Economic Growth. More Welfare through Less Production? Amsterdam (etc.): North-Holland Publishing Company.

Humphrey, M., 2002. Preservation Versus the People? Nature, Humanity, and Political Philosophy. Oxford (etc.): Oxford University Press.

Humphrey, M., 2003. Nonbasic Environmental Goods and Social Justice. In: D.A. Bell and A. de-Shalit, eds. Forms of Justice: Critical Perspectives on David Miller's Political Philosophy. Lanham: Rowman and Littlefield, 331–346.

Hyams, K., 2009. A Just Response to Climate Change: Personal Carbon Allowances and the Normal-Functioning Approach. Journal of Social Philosophy, 40 (2), 237–256.

Illich, I., 1973. Tools for conviviality. New York (etc.): Harper & Row.

IPCC (Intergovernmental Panel on Climate Change), 2014. Climate Change 2014: Synthesis Report. Contribution of Working Groups I, II and III to the Fifth Assessment Report of the Intergovernmental Panel on Climate Change. Geneva: IPCC.

Jackson, T., 2005. Live Better by Consuming Less? Is There a 'Double Dividend' in Sustainable Consumption? Journal of Industrial Ecology, 9 (1-2), 19–36.

Jackson, T., 2007. Sustainable Consumption. In: G. Atkinson, S. Dietz, and E. Neumayer, eds. Handbook of Sustainable Development. Cheltenham (etc.): Edward Elgar, 254–268.

Jackson, T., 2009. Prosperity without growth. London (etc.): Earthscan.

Jaeger, C.C., Paroussos, L., Mangalagiu, D., Kupers, R., Mandel, A., Tàbara, J.D., Meißner, F., and Lass, W., 2011. A New Growth Path for Europe. Generating Prosperity and Jobs in the Low-Carbon Economy. Final Report. A study commissioned by the German Federal Ministry for the Environment, Nature Conservation and Nuclear Safety. Potsdam: European Climate Forum e.V.

Jänicke, M., 1996. Erfolgsbedingungen von Umweltpolitik. In: M. Jänicke, ed. Umweltpolitik der Industrieländer. Entwicklung, Bilanz, Erfolgsbedingungen. Berlin: Edition Sigma, 9–28.

Jänicke, M., 2003. Umweltpolitik. In: M. Jänicke, P. Kunig, and M. Stitzel, eds. Umweltpolitik. Politik, Recht und Management des Umweltschutzes in Staat und Unternehmen. Bonn: Dietz, 30–157.

Jänicke, M., 2009. On ecological and political modernization. In: A.P.J. Mol, D.A. Sonnenfeld, and G. Spaargaren, eds. The ecological modernisation reader: environmental reform in theory and practice. London, New York: Routledge, 28–41.

Jänicke, M., 2011. 'Green Growth': From a growing eco-industry to a sustainable economy. Berlin: Freie Universität, Environmental Policy Research Centre, No. FFU-Report 09-2011.

Jänicke, M. and Jacob, K., 2009. A Third Industrial Revolution? Solutions to the crisis of resource-intensive growth. Berlin: Freie Universität, Environmental Policy Research Centre, FFU-report No. 02-2009.

Jänicke, M. and Jörgens, H., 2009. New Approaches to Environmental Governance. In: A.P.J. Mol, D.A. Sonnenfeld, and G. Spaargaren, eds. The ecological modernisation reader: environmental reform in theory and practice. London, New York: Routledge, 156–189.

Jenkins, J., Nordhaus, T., and Shellenberger, M., 2011. Energy Emergence: Rebound & Backfire as emergent Phenomena. Oakland (CA): Breakthrough Institute.

Jensen, P.D., Basson, L., and Leach, M., 2011. Reinterpreting Industrial Ecology. Journal of Industrial Ecology, 15 (5), 680–692.

Johnson, B.L., 2003. Ethical obligations in a tragedy of the commons. Environmental Values, 12 (3), 271–287.

Jonas, H., 1979. Das Prinzip Verantwortung: Versuch einer Ethik für die technologische Zivilisation. Frankfurt am Main: Insel-Verlag.

Kassenboehmer, S.C. and Schmidt, C.M., 2011. Beyond GDP and Back: What is the Value-Added by Additional Components of Welfare Measurement? Berlin.

Kates, R.W., Clark, W.C., Corell, R., Hall, J.M., Jaeger, C.C., Lowe, I., McCarthy, J.J., Schellnhuber, H.J., Bolin, B., Dickson, N.M., Faucheux, S., Gallopin, G.C., Grübler, A., Huntley, B., Jäger, J., Jodha, N.S., Kasperson, R.E., Mabogunje, A., Matson, P., Mooney, H., III, B.M., O'Riordan, T., and Svedin, U., 2001. Sustainability Science. Science, 292 (5517), 641–642.

Kates, R.W., Parris, T.M., and Leiserowitz, A.A., 2005. What Is Sustainable Development? Goals, Indicators, Values, and Practice. Environment, 47 (3), 9–21.

Keesing, F., Belden, L.K., Daszak, P., Dobson, A., Harvell, C.D., Holt, R.D., Hudson, P., Jolles, A., Jones, K.E., Mitchell, C.E., Myers, S.S., Bogich, T., and Ostfeld, R.S., 2010. Impacts of biodiversity on the emergence and transmission of infectious diseases. Nature, 468, 647–652.

Kelly, P., 2005. Liberalism. Cambridge (etc.): Polity Press.

Kemp, R., 1997. Environmental Policy and Technical Change. A Comparison of the Technological Impact of Policy Instruments. Cheltenham (etc.): Edward Elgar.

Kemp, R., 2010. Sustainable technologies do not exist! Ekonomiaz, 3 (75), 2–17.

Kernohan, A., 1993. Accumulative Harms and the Interpretation of the Harm Principle. Social Theory and Practice, 19 (1), 51–72.

Kernohan, A., 1998. Liberalism, Equality, and Cultural Oppression. Cambridge (etc.): Cambridge University Press.

Kernohan, A., 2000. Individual Acts and Accumulative Consequences. Philosophical Studies, 97 (3), 343–366.

Keynes, J.M., 1930. Economic Possibilities for our Grandchildren.

Kleinhückelkotten, S., 2005. Suffizienz und Lebensstile : Ansätze für eine milieuorientierte Nachhaltigkeitskommunikation. Berlin: Berliner Wissenschafts-Verlag.

Kollmuss, A. and Agyeman, J., 2002. Mind the Gap: why do people act environmentally and what are the barriers to pro-environmental behavior? Environmental Education Research, 8 (3), 239–260.

Kopfmüller, J., 2006. Das integrative Konzept nachhaltiger Entwicklung: Motivation, Architektur, Perspektiven. In: J. Kopfmüller, ed. Ein Konzept auf dem Prüfstand—Das integrative Nachhaltigkeitskonzept in der Forschungspraxis. Berlin: Edition Sigma, 23–37.

Kopfmüller, J., Brandl, V., Jörissen, J., Paetau, M., Banse, G., Coenen, R., and Grunwald, A., 2001. Nachhaltige Entwicklung integrativ betrachtet: konstitutive Elemente, Regeln, Indikatoren. Berlin: Edition Sigma.

Kraemer, K., 1997. Nachhaltigkeit durch Konsumverzicht? „Sustainable Development"—eine soziologische Betrachtung. Zeitschrift für angewandte Umweltforschung, 10 (2), 198–209.

Kraftfahrt-Bundesamt, 2015. Pressemitteilung Nr. 5/2015: Der Fahrzeugbestand am 1. Januar 2015.

Krausmann, F., Gingrich, S., Eisenmenger, N., Erb, K.-H., Haberl, H., and Fischer-Kowalski, M., 2009. Growth in global materials use, GDP and population during the 20th century. Ecological Economics, 68 (10), 2696–2705.

Krebs, A., 2001. Wieviel Natur schulden wir der Zukunft? Eine Kritik am zukunftsethischen Egalitarismus. In: D. Birnbacher and G. Brudermüller, eds. Zukunftsverantwortung und Generationensolidarität. Würzburg: Königshausen & Neumann, 157–183.

Krebs, A., 2003. Warum Gerechtigkeit nicht als Gleichheit zu begreifen ist. Deutsche Zeitschrift für Philosophie, 51 (2), 235–253.

Krebs, A., ed., 2000. Gleichheit oder Gerechtigkeit. Texte der neuen Egalitarismuskritik. Frankfurt a. M.: Suhrkamp.

Kroes, P., 1998. Philosophy of Technology. In: E. Craig, ed. Routledge Encyclopedia of Philosophy. London, New York: Routledge, 284–288.

Kuhn, K., 2006. Zur kulturellen Dimension nachhaltiger Entwicklung—Eine metatheoretische und diskursanalytische Bestandsaufnahme. INFU-Diskussionsbeiträge 28/06. Lüneburg.

Kuhnimhof, T., Armoogum, J., Buehler, R., Dargay, J., Denstadli, J.M., and Yamamoto, T., 2012. Men Shape a Downward Trend in Car Use among Young Adults—Evidence from Six Industrialized Countries. Transport Reviews, 32 (6), 761–779.

Kunig, P., 2003. Umweltrecht. In: M. Jänicke, P. Kunig, and M. Stitzel, eds. Umweltpolitik. Politik, Recht und Management des Umweltschutzes in Staat und Unternehmen. Bonn: Dietz, 159–288.

Kurz, R., 2007. Wirtschaftswachstum als Ausweg aus der ökologisch-sozialen Krise? In: S. Rudolph, ed. Wachstum, Wachstum über alles? Ein ökonomisches Leitbild auf dem Prüfstand von Umwelt und Gerechtigkeit. Marburg: Metropolis, 17–30.

Kutz, C., 2000. Complicity: Ethics and Law for a Collective Age. Cambridge: Cambridge University Press.

Kymlicka, W., 1989. Liberal Individualism and Liberal Neutrality. Ethics, 99, 883–905.

Langhelle, O., 2000. Why ecological modernization and sustainable development should not be conflated. Journal of Environmental Policy & Planning, 2 (4), 303–322.

Larmore, C., 2003. Patterns of Moral Complexity. Reprinted excerpts from the original by Cambridge University Press (1987). In: S. Wall and G. Klosko, eds. Perfectionism and Neutrality: Essays in Liberal Theory. Lanham (etc.): Rowman and Littlefield, 47–60.

Latouche, S., 2009. Farewell to growth. Cambridge (etc.): Polity Press.

Layard, R., 2005. Die glückliche Gesellschaft. Frankfurt/Main, New York: Campus-Verlag.

Leahy, E., Lyons, S., and Tol, R.S.J., 2010. An estimate of the number of vegetarians in the world. ESRI working paper No. 340.

Leipert, C., 1991. Die andere Seite der Wachstumsmedaille: Ökologische und soziale Folgekosten des Wirtschaftens in der Industriegesellschaft. In: H. Diefenbacher and S. Habicht-Erenler, eds. Wachstum und Wohlstand: neuere Konzepte zur Erfassung der Sozial- und Umweltverträglichkeit. Marburg: Metropolis, 21–41.

Lerch, A. and Nutzinger, H.G., 2002. Sustainability: Economic Approaches and Ethical Implications. Journal of Economic and Social Policy, 6 (2).

Lerch, A., 2006. Kann man Nachhaltigkeit ökonomisch begründen? In: F. Ekardt, ed. Generationenengerechtigkeit und Zukunftsfähigkeit. Philosophische, juristische, ökonomische, politologische und theologische Neuansätze. Hamburg: Lit Verlag, 108–128.

Levin, S.A. and Clark, W.C., 2010. Toward a Science of Sustainability: Report from Toward a Science of Sustainability Conference November 29, 2009 - December 2, 2009. Warrenton, Virginia: The National Science Foundation.

Lifset, R., 1998. Setting the Boundaries? Journal of Industrial Ecology, 2 (2), 1–2.

Lin, A.C., 2006. The Unifying Role of Harm in Environmental Law. Wisconsin Law Review, 897–985.

Linklater, A., 2011. The Problem of Harm in World Politics—Theoretical Investigations. Cambridge (etc.): Cambridge University Press.

Linz, M. and Kristof, K., 2007. Suffizienz, nicht Wachstum, ist der Schlüssel für mehr Lebensqualität. In: S. Rudolph, ed. Wachstum, Wachstum über alles? Ein ökonomisches Leitbild auf dem Prüfstand von Umwelt und Gerechtigkeit. Marburg: Metropolis, 177–191.

Linz, M. and Scherhorn, G., 2011. Für eine Politik der Energie-Suffizienz. Wuppertal: Wuppertal Institut.

Linz, M., 2002. Warum Suffizienz unentbehrlich ist. In: M. Linz, P. Bartelmus, P. Hennicke, R. Jungkeit, W. Sachs, G. Scherhorn, G. Wilke, and U. v. Winterfeld, eds. Von nichts zu viel—Suffizienz gehört zur Zukunftsfähigkeit. 7–14.

Linz, M., 2004. Weder Mangel noch Übermaß. Über Suffizienz und Suffizienzforschung. Wuppertal Paper Nr. 145.

Linz, M., 2006. Was wird dann aus der Wirtschaft? Über Suffizienz, Wirtschaftswachstum und Arbeitslosigkeit. Wuppertal Paper Nr. 157.

Linz, M., 2012 a. Suffizienz als Überlebensfrage. Umwelt aktuell, 2–3.

Linz, M., 2012 b. Weder Mangel noch Übermaß—Warum Suffizienz unentbehrlich ist. München: oekom Verlag.

Linz, M., 2015. Suffizienz als politische Praxis—Ein Katalog. Wuppertal, Wuppertal Spezial 49.

Linz, M., 2017. Wie Suffizienzpolitiken gelingen. Wuppertal, Wuppertal Spezial 52.

Linz, M., Bartelmus, P., Hennicke, P., Jungkeit, R., Sachs, W., Scherhorn, G., Wilke, G., and Winterfeld, U. v., 2002. Von nichts zu viel—Suffizienz gehört zur Zukunftsfähigkeit. Wuppertal Paper Nr. 125.

Luks, F., 1999. Der steady-state als Grundlage eines sustainable development. Hamburg: Hamburger Universität für Wirtschaft und Politik.

Luks, F., 2001. Die Zukunft des Wachstums: Theoriegeschichte, Nachhaltigkeit und die Perspektiven einer neuen Wirtschaft. Marburg: Metropolis.

Luks, F., 2007. Jenseits des Ökonomischen das Nachhaltige suchen. Ökologisches Wirtschaften, (1/2007), 27–29.

MacCallum, G.C., 1967. Negative and Positive Freedom. The Philosophical Review, 76 (3), 312–334.

Macmillan, A., Connor, J., Witten, K., Kearns, R., Rees, D., and Woodward, A., 2014. The Societal Costs and Benefits of Commuter Bicycling: Simulating the Effects of Specific Policies Using System Dynamics Modeling. Environmental Health Perspectives, 122 (4), 335–344.

Maniates, M., 2002. Individualization: Plant a tree, buy a bike, save the world? In: T. Princen, M. Maniates, and K. Conca, eds. Confronting Consumption. Cambridge (Mass.): MIT Press, 43–66.

Margalit, A., 1997. Decent Equality and Freedom: A Postscript. Social Research, 64 (1), 147–160.

Martin, R., 1998. Rights. In: E. Craig, ed. Routledge Encyclopedia of Philosophy. London, New York: Routledge, 325–331.

Martínez-Alier, J., Pascual, U., Vivien, F.-D., and Zaccai, E., 2010. Sustainable degrowth: Mapping the context, criticisms and future prospects of an emergent paradigm. Ecological Economics, 69 (9), 1741–1747.

Max-Neef, M.A., Elizalde, A., and Hopenhayn, M., 1991. Development and Human Needs. In: M.A. Max-Neef, ed. Human Scale Development: Conception, Application, and Further Reflections. New York, London: The Apex Press, 13–54.

Maxwell, D., Owen, P., McAndrew, L., Muehmel, K., and Neubauer, A., 2011. Addressing the Rebound Effect. A report for the European Commission DG Environment.

Mayntz, R., 1987. Politische Steuerung und gesellschaftliche Steuerungsprobleme: Anmerkungen zu einem theoretischen Paradigma. In: Jahrbuch zur Staats- und Verwaltungswissenschaft. Baden-Baden: Nomos, 89–110.

Mayntz, R., 2004. Governance Theory als fortentwickelte Steuerungstheorie?

McDonough, W. and Braungart, M., 1998. The NEXT Industrial Revolution. The Atlantic Monthly, 282 (4), 82–92.

McDonough, W. and Braungart, M., 2002. Cradle to Cradle: Remaking the Way We Make Things. New York: North Point Press.

McDonough, W., 1998. A Boat for Thoreau: A Discourse on Ecology, Ethics, and the Making of Things. In: L. Westra and P.H. Werhane, eds. The Business of Consumption: Environmental Ethics and the Global Economy. Lanham (etc.): Rowman and Littlefield, 297–317.

McLaren, D., 2003. Environmental Space, Equity and the Ecological Debt. In: J. Agyeman, R.D. Bullard, and B. Evans, eds. Just sustainabilities: development in an unequal world. Cambridge (Mass.): MIT Press, 19–37.

McLean, I. and McMillan, A., eds., 2009. The Concise Oxford Dictionary of Politics. Oxford, New York: Oxford University Press.

McNeill, J., 2005. Blue Planet—Die Geschichte der Umwelt im 20. Jahrhundert. Bonn: Bundeszentrale für politische Bildung.

Meadows, D.H., Randers, J., and Meadows, D.L., 2004. Limits to Growth: The 30-year Update. White River Junction: Chelsea Green Publishing Company.

Meadows, D.L., Meadows, D.H., Randers, J., and Behrens, W.W., 1972. The Limits to Growth: A Report for the Club of Rome's Project on the Predicament of Mankind. New York: Universe Books.

Meyer, J.M., 2011. We have never been liberal: the environmentalist turn to liberalism and the possibilities for social criticism. Environmental Politics, 20 (3), 356–373.

Meyer, L., 2008. Intergenerational Justice. The Stanford Encyclopedia of Philosophy. http://plato.stanford.edu/archives/spr2010/entries/justice-intergenerational/ (last accessed October 2012).

Meyer, L.H., 2009. Intergenerationelle Suffizienzgerechtigkeit. In: N. Goldschmidt, ed. Generationengerechtigkeit: Ordnungsökonomische Konzepte. Tübingen: Mohr Siebeck, 281–322.

Michelsen, G. and Adomßent, M., 2014. Nachhaltige Entwicklung: Hintergründe und Zusammenhänge. In: H. Heinrichs and G. Michelsen, eds. Nachhaltigkeitswissenschaften. Berlin, Heidelberg: Springer, 3–59.

Midden, C.J.H., Kaiser, F.G., and McCalley, L.T., 2007. Technology's Four Roles in Understanding Individuals' Conservation of Natural Resources. Journal of Social Issues, 63 (1), 155–174.

Miegel, M., 2010. Exit. Berlin: Propyläen.

Mill, J.S., 1909 [1848]. Principles of Political Economy with some of their Applications to Social Philosophy. Library of Economics and Liberty.

Mill, J.S., 2009 [1859]. On Liberty / Über die Freiheit. Stuttgart: Reclam.

Millennium Ecosystem Assessment, 2005. Ecosystems and Human Well-being: Synthesis. Washington, DC.: Island Press.

Miller, D., 1976. Social Justice. Oxford: Clarendon Press.

Miller, D., 1999 a. Social Justice and Environmental Goods. In: A. Dobson, ed. Fairness and Futurity—Essays on Environmental Sustainability and Social Justice. Oxford, New York: Oxford University Press, 151–172.

Miller, D., 1999 b. Justice and Global Inequality. In: A. Hurrell and N. Woods, eds. Inequality, Globalization, and World Politics. Oxford, New York: Oxford University Press, 187–210.

Miller, D., 2004. Justice, democracy and public goods. In: K. Dowding, R.E. Goodin, and C. Pateman, eds. Justice and Democracy: Essays for Brian Barry. Cambridge: Cambridge University Press, 127–149.

Miller, D., 2008. Global Justice and Climate Change: How Should Responsibilities Be Distributed? The Tanner Lectures on Human Values, 119–156. http://www.tannerlectures.utah.edu/lectures/documents/Miller_08.pdf (last accessed March 2013).

Miller, D., 2012. Consumption and its consequences. Cambridge (etc.): Polity Press.

Mol, A.P.J. and Jänicke, M., 2009. The origins and theoretical foundations of ecological modernisation theory. In: A.P.J. Mol, D.A. Sonnenfeld, and G. Spaargaren, eds. The ecological modernisation reader: environmental reform in theory and practice. London, New York: Routledge, 17–27.

Mol, A.P.J. and Spaargaren, G., 2000. Ecological Modernisation Theory in Debate: A Review. In: A.P.J. Mol and D.A. Sonnenfeld, eds. Ecological Modernisation around the world: perspectives and critical debates. London: Frank Cass Publishers, 17–49.

Mol, A.P.J., Sonnenfeld, D.A., and Spaargaren, G., eds., 2009. The ecological modernisation reader: environmental reform in theory and practice. London, New York: Routledge.

Molenbroek, E., Smith, M., Groenenberg, H., Waide, P., Attali, S., Fischer, C., Krivošik, J., Fonseca, P., Santos, B., and Fong, J., 2014. Final technical report: Evaluation of the Energy Labelling Directive and specific aspects of the Ecodesign Directive. Utrecht: Ecofys, No. ENER/C3/2012 - 523.

Morley, B., 2012. Empirical Evidence on the Effectiveness of Environmental Taxes. Applied Economics Letters, 19 (18), 1817–1820.

Müller, A., 2009. Sufficiency—does energy consumption become a moral issue? In: C. Broussous and C. Jover, eds. Act! Innovate! Deliver! Reducing Energy Demand Sustainably. ECEEE 2009 Summer Study, Stockholm, 83–90.

Müller, O.L., 2011. Eine faire Lösung des Klimaproblems. Information Philosophie, 39 (2), 91–98.

Muraca, B., 2009. Nachhaltigkeit ohne Wachstum? Auf dem Weg zur Décroissance. In: T. v. Egan-Krieger, J. Schultz, P.P. Thapa, and L. Voget, eds. Die Greifswalder Theorie starker Nachhaltigkeit. Ausbau, Anwendung und Kritik. Marburg: Metropolis, 241–259.

Nagel, T., 2005. The Problem of Global Justice. Philosophy & Public Affairs, 33 (2), 113–147.

Neumayer, E., 1999. Weak versus strong sustainability: exploring the limits of two opposing paradigms. Cheltenham (etc.): Elgar.

Neumayer, E., 2000. Nicht Schuldig! Politische Ökologie, (66), 20–23.

Newig, J., 2007. Symbolic environmental legislation and societal self-deception. Environmental Politics, 16 (2), 276–296.

Nida-Rümelin, J., 2006. Demokratie und Wahrheit. München: Verlag C.H.Beck.

Nida-Rümelin, J., 2007. Politische Verantwortung. In: L. Heidbrink and A. Hirsch, eds. Staat ohne Verantwortung? Zum Wandel der Aufgaben von Staat und Politik. Frankfurt, New York: Campus Verlag, 55–85.

Nozick, R., 1969. Coercion. In: S. Morgenbesser, P. Suppes, and M. White, eds. Philosophy, Science, and Method: Essays in Honor of Ernest Nagel. New York: St. Martin's Press, 440–472.

Nozick, R., 1974. Anarchy, State, and Utopia. Oxford: Basil Blackwell.

Nussbaum, M.C., 2011. Creating Capabilities. The Human Development Approach. Cambridge, London: The Belknap Press of Harvard University Press.

Nutzinger, H.G., 2006. Starke versus schwache Nachhaltigkeit. In: F. Ekardt, ed. Generationengerechtigkeit und Zukunftsfähigkeit. Philosophische, juristische, ökonomische, politologische und theologische Neuansätze. Hamburg: Lit Verlag, 186–200.

O'Neill, D., Dietz, R., and Jones, N., 2010. Enough is Enough: Ideas for a Sustainable Economy in a World of Finite Resources. The Report of the Steady State Economy Conference. Leeds, UK: Center for the Advancement of the Steady State Economy and Economic Justice for All.

O'Neill, J., 2011. The overshadowing of needs. In: F. Rauschmayer, I. Omann, and J. Frühmann, eds. Sustainable Development: Capabilities, needs, and well-being. Oxon, New York: Routledge, 25–42.

OECD (Organisation for Economic Co-operation and Development), 2002 a. Towards Sustainable Household Consumption? Trends and Policies in OECD Countries. OECD Policy Brief.

OECD (Organisation for Economic Co-operation and Development), 2002 b. Policy Case Study Series—Policies to Promote Sustainable Consumption: An Overview. ENV/EPOC/WPNEP(2001)18/FINAL.

OECD (Organisation for Economic Co-operation and Development), 2003. Voluntary Approaches for Environmental Policy: Effectiveness, Efficiency and Usage in Policy Mixes. Paris: OECD Publishing.

OECD (Organisation for Economic Co-operation and Development), 2008. Promoting Sustainable Consumption: Good Practices in OECD Countries. Paris: OECD Publishing.

OECD (Organisation for Economic Co-operation and Development), 2011 a. Greening Household Behaviour: The Role of Public Policy. OECD Publishing.

OECD (Organisation for Economic Co-operation and Development), 2011 b. Towards Green Growth. OECD Publishing.

oekom, ed., 2013. Vom rechten Maß—Suffizienz als Schlüssel zu mehr Lebensglück und Umweltschutz. Politische Ökologie, 134.

Ölander, F. and Thøgersen, J., 2014. Informing Versus Nudging in Environmental Policy. Journal of Consumer Policy, 37, 341–356.

Ostrom, E., 1990. Governing the Commons. The Evolution of Institutions for Collective Action. Cambridge (etc.): Cambridge University Press.

Ott, K. and Döring, R., 2004. Theorie und Praxis starker Nachhaltigkeit. Marburg: Metropolis.

Ott, K. and Döring, R., 2007. Soziale Nachhaltigkeit: Suffizienz zwischen Lebensstilen und politischer Ökonomie. In: F. Beckenbach, U. Hampicke, C. Leipert, G. Meran, J. Minsch, H.G. Nutzinger, R. Pfriem, J. Weimann, F. Wirl, and U. Witt, eds. Soziale Nachhaltigkeit. Marburg: Metropolis, 35–71.

Ott, K. and Voget, L., 2007. Suffizienz: Umweltethik und Lebensstilfragen. http://www.boell.de/downloads/oekologie/Vortrag_Suffizienz_Ende_Konrad_Ott_100607.pdf (last accessed October 2012).

Ott, K., 2004. Theoriebildung statt Definitionswirrwarr. GAIA, 13 (1), 38–39.

Oxfam, 2008. Another Inconvenient Truth: How biofuel policies are deepening poverty and accelerating climate change. Oxford.

Pacala, S. and Socolow, R., 2004. Stabilization Wedges: Solving the Climate Problem for the Next 50 Years with Current Technologies. Science, 305 (5686), 968–972.

Paech, N., 2009. Unternehmerische Nachhaltigkeit, Suffizienz und Zeitknappheit. In: I. Antoni-Komar, M. Beermann, C. Lautermann, J. Müller, N. Paech, H. Schattke, U. Schneidewind, and R. Schulz, eds. Neue Konzepte der Ökonomik: Unternehmen zwischen Nachhaltigkeit, Kultur und Ethik. Festschrift für Reinhard Pfriem zum 60. Geburtstag. Marburg: Metropolis, 61–83.

Paech, N., 2011. Adiós Konsumwohlstand: Vom Desaster der Nachhaltigkeitskommunikation und den Möglichkeiten der Suffizienz. In: L. Heidbrink, I. Schmidt, and B. Ahaus, eds. Die Verantwortung des Konsumenten. Über das Verhältnis von Markt, Moral und Konsum. Frankfurt, New York: Campus Verlag, 285–304.

Page, E.A., 2007 a. Justice Between Generations: Investigating a Sufficientarian Approach. Journal of Global Ethics, 3 (1), 3–20.

Page, E.A., 2007 b. Intergenerational justice of what: Welfare, resources or capabilities? Environmental Politics, 16 (3), 453–469.

Paqué, K.-H., 2010. Wachstum! Die Zukunft des globalen Kapitalismus. München: Carl Hanser Verlag.

Parfit, D., 1982. Future Generations: Further Problems. Philosophy & Public Affairs, 11 (2), 113–172.

Parfit, D., 1997. Equality and Priority. Ratio, 10 (3), 202–221.

Parris, T.M. and Kates, R.W., 2003. Characterizing and Measuring Sustainable Development. Annual Review of Environment and Resources, 28, 559–586.

Parry, I.W.H., 2002. Comparing the efficiency of alternative policies for reducing traffic congestion. Journal of Public Economics, 85 (3), 333–362.

Pearce, D., 1988. Economics, equity and sustainable development. Futures, 20 (6), 598–605.

Pearce, D., 1990. Economics and the Global Environmental Challenge. Journal of International Studies, 19 (3), 365–387.

Pearce, D., 1993. Blueprint 3: Measuring Sustainable Development. London: Earthscan.

Pearce, D., 1997. Substitution and sustainability: some reflections on Georgescu-Roegen. Ecological Economics, 22, 295–297.

Pearce, D.W. and Turner, R.K., 1990. Economics of Natural Resources and the Environment. New York (etc.): Harvester Wheatsheaf.

Pellizzoni, L., 2010. Environmental Knowledge and Deliberative Democracy. In: M. Groß and H. Heinrichs, eds. Environmental Sociology European Perspectives and Interdisciplinary Challenges. New York (etc.): Springer, 159–182.

Peters, G.P. and Hertwich, E.G., 2006. The Importance of Imports for Household Environmental Impacts. Journal of Industrial Ecology, 10 (3), 89–109.

Peters, G.P., Minx, J.C., Weber, C.L., and Edenhofer, O., 2011. Growth in emission transfers via international trade from 1990 to 2008. Proceedings of the National Academy of Sciences, 108 (21), 8903–8908.

Petersen, T. and Schiller, J., 2011. Politische Verantwortung für Nachhaltigkeit und Konsumentensouveränität. GAIA, 20 (3), 157–161.

Petschow, U., 2007. Nachhaltigkeit und Innovation im Brundtland-Bericht—Harmonische Konflikte um die Zukunft des Wirtschaftens. Ökologisches Wirtschaften, (1/2007), 25–26.

Pogge, T., 2005. Real World Justice. Ethics, 9, 29–53.

Pogge, T.W., 2004. 'Assisting' the global poor. In: D.K. Chatterjee, ed. The ethics of assistance: Morality and the distant needy. Cambridge (etc.): Cambridge University Press, 260–288.

Princen, T., 1999. Consumption and environment: some conceptual issues. Ecological Economics, 31 (3), 347–363.

Princen, T., 2002. Consumption and its externalities: where economy meets ecology. In: T. Princen, M. Maniates, and K. Conca, eds. Confronting Consumption. Cambridge (Mass.): MIT Press, 23–42.

Princen, T., 2005. The logic of sufficiency. Cambridge (Mass.) (etc.): MIT Press.

Princen, T., Maniates, M., and Conca, K., eds., 2002. Confronting Consumption. Cambridge (Mass.): MIT Press.

Radkau, J., 2008. Technik in Deutschland vom 18. Jahrhundert bis heute. Frankfurt, New York: Campus Verlag.

Radkau, J., 2011. Die Ära der Ökologie—Eine Weltgeschichte. München: Verlag C.H.Beck.

Rauschmayer, F., Omann, I., and Frühmann, J., 2011. Needs, capabilities and quality of life—Refocusing sustainable development. In: F. Rauschmayer, I. Omann, and J. Frühmann, eds. Sustainable Development: Capabilities, needs, and well-being. Oxon, New York: Routledge, 1–24.

Rawls, J., 1972. A Theory of Justice. London (etc.): Oxford University Press.

Rawls, J., 1998. Politischer Liberalismus. Frankfurt a. M.: Suhrkamp.

Rawls, J., 1999 a. A Theory of Justice—Revised Edition. Oxford, New York: Oxford University Press.

Rawls, J., 1999 b. The Law of Peoples. Cambridge, London: Harvard University Press.

Raworth, K., 2012. A safe and just space for humanity: Can we live within the Doughnut? Oxfam.

Raz, J., 1986. The Morality of Freedom. Oxford: Clarendon Press.

Rees, W.E. and Westra, L., 2003. When consumption does violence: can there be sustainability and environmental justice in a resource-limited world? In: J. Agyeman, R.D. Bullard, and B. Evans, eds. Just sustainabilities: development in an unequal world. Cambridge (Mass.): MIT Press, 99–124.

Reichel, A., Goll, F., and Scheiber, L., 2009. Linking Sufficiency and Business: Utility Systems Engineering in Producer-Consumer Networks. Presented at the Academy of Management Annual Meeting.

Reisch, L.A. and Hagen, K., 2011. Kann der Konsumwandel gelingen? Chancen und Grenzen einer verhaltensökonomisch basierten sozialen Regulierung. In: L. Heidbrink, I. Schmidt, and B. Ahaus, eds. Die Verantwortung des Konsumenten. Über das Verhältnis von Markt, Moral und Konsum. Frankfurt, New York: Campus Verlag, 221–243.

Reisch, L.A. and Sandrini, J., 2015. Nudging in der Verbraucherpolitik. Ansätze verhaltensbasierter Regulierung. Baden-Baden: Nomos.

Renn, O., 2008. Risk Governance: Coping with Uncertainty in a Complex World. London (etc.): Earthscan.

Reusswig, F. and Battaglini, A., 2006. Lifestyle Dynamics as a Catalyst of a Sustainable Energy Transition. Presented at the KyotoPlus, Berlin.

Rietveld, P. and Daniel, V., 2004. Determinants of Bicycle Use: Do Municipal Policies Matter? Transportation Research Part A, 38, 531–550.

Rifkin, J., 2011. The Third Industrial Revolution: How Lateral Power is Transforming Energy, the Economy, and the World. Basingstoke: Palgrave Macmillan.

Risse, M., 2012. On Global Justice. Princeton, Oxford: Princeton University Press.

Röckseisen, S., 1998. Kriterien zur Beurteilung der Leistungsfähigkeit umweltpolitischer Instrumente. In: M. Kotulla, H. Ristau, and U. Smeddinck, eds. Heidelberg: Physica, 171–183.

Rockström, J., Steffen, W., Noone, K., Persson, Å., Chapin, F.S., Lambin, E.F., Lenton, T.M., Scheffer, M., Folke, C., Schellnhuber, H.J., Nykvist, B., Wit, C.A. de, Hughes, T., Leeuw, S. van der, Rodhe, H., Sörlin, S., Snyder, P.K., Costanza, R., Svedin, U., Falkenmark, M., Karlberg, L., Corell, R.W., Fabry, V.J., Hansen, J., Walker, B., Liverman, D., Richardson, K., Crutzen, P., and Foley, J.A., 2009 a. Planetary boundaries: exploring the safe operating space for humanity. Ecology and Society, 14 (2), 32.

Rockström, J., Steffen, W., Noone, K., Persson, Å., Chapin, F.S., Lambin, E.F., Lenton, T.M., Scheffer, M., Folke, C., Schellnhuber, H.J., Nykvist, B., Wit, C.A. de, Hughes, T., Leeuw, S. van der, Rodhe, H., Sörlin, S., Snyder, P.K., Costanza, R., Svedin, U., Falkenmark, M., Karlberg, L., Corell, R.W., Fabry, V.J., Hansen, J., Walker, B., Liverman, D., Richardson, K., Crutzen, P., and Foley, J.A., 2009 b. A safe operating space for humanity. Nature, 461, 472–475.

Røpke, I., 1999. The dynamics of willingness to consume. Ecological Economics, 28 (3), 399–420.

Røpke, I., 2010. Konsum: Der Kern des Wachstumsmotors. In: I. Seidl and A. Zahrnt, eds. Postwachstumsgesellschaft. Konzepte für die Zukunft. Marburg: Metropolis, 103–115.

Rosa, H., 2011. Über die Verwechslung von Kauf und Konsum: Paradoxien der spätmodernen Konsumkultur. In: L. Heidbrink, I. Schmidt, and B. Ahaus, eds. Die Verantwortung des Konsumenten. Über das Verhältnis von Markt, Moral und Konsum. Frankfurt, New York: Campus Verlag, 115–132.

Rosenberg, A., 1995. Equality, Sufficiency, and Opportunity in the Just Society. Social Philosophy and Policy, 12 (2), 54–71.

Rubik, F., Scholl, G., Biedenkopf, K., Kalimo, H., Mohaupt, F., Söebech, Ó., Stø, E., Strandbakken, P., and Turnheim, B., 2009. Innovative Approaches in European Sustainable Consumption Policies. Assessing the potential of various instruments for sustainable consumption practises and greening of the market (ASCEE). Berlin: IÖW.

Rückert-John, J., Jaeger-Erben, M., Schäfer, M., Aderhold, J., and John, R., 2013. Soziale Innovationen für nachhaltigen Konsum: Kriterien zur Analyse und Systematisierung. Berlin: Institut für Sozialinnovation.

Rudolph, S., ed., 2007. Wachstum, Wachstum über alles? Ein ökonomisches Leitbild auf dem Prüfstand von Umwelt und Gerechtigkeit. Marburg: Metropolis.

Sachs, J.D., 2015. The Age of Sustainable Development. New York (u.a.): Columbia University Press.

Sachs, W., 1993. Die vier E's. Politische Ökologie (Special Sept./Okt. '93), 69–72.

Sachs, W., 2009. Anders! Weniger! Besser! Konturen eines zukunftsfähigen Wirtschaftsmodells. Politische Ökologie, (118), 37–40.

Sagoff, M., 1988. The economy of the earth: Philosophy, law, and the environment. Cambridge: Cambridge University Press.

Sagoff, M., 1997. Do we consume too much? The Atlantic Monthly, 279 (6), 80–96.

Sanne, C., 2002. Willing consumers—or locked-in? Policies for a sustainable consumption. Ecological Economics, 42 (1-2), 273–287.

Satz, D., 2003. International Economic Justice. In: H. LaFollette, ed. The Oxford Handbook of Practical Ethics. Oxford: Oxford University Press, 620–642.

Saward, M., 1993. Green democracy? In: A. Dobson and P. Lucardie, eds. The politics of nature: explorations in green political theory. London, New York: Routledge, 63–80.

Saward, M., 1996. Must democrats be environmentalists? In: B. Doherty and M. de Geus, eds. Democracy and Green Political Thought—Sustainability, rights and citizenship. London, New York: Routledge, 79–96.

Scanlon, T.M., 2003. The Diversity of Objections to Inequality. In: T.M. Scanlon, ed. The Difficulty of Tolerance: Essays in Political Philosophy. Cambridge (UK) (etc.): Cambridge University Press, 202–218.

Schefczyk, M., 2011. Das John-Stuart-Mill-Problem. In: F. Höntzsch, ed. John Stuart Mill und der sozialliberale Staatsbegriff. Stuttgart: Franz Steiner Verlag, 27–42.

Scherhorn, G., 1994. Die Unersättlichkeit der Bedürfnisse und der kalte Stern der Knappheit. In: B. Biervert and M. Held, eds. Das Naturverständnis der Ökonomik. Beiträge zur Ethikdebatte in den Wirtschaftswissenschaften. Frankfurt, New York: Campus Verlag, 224–238.

Scherhorn, G., 2002. Die Logik der Suffizienz. In: M. Linz, P. Bartelmus, P. Hennicke, R. Jungkeit, W. Sachs, G. Scherhorn, G. Wilke, and U. v. Winterfeld, eds. Von nichts zu viel—Suffizienz gehört zur Zukunftsfähigkeit. 15–26.

Scherhorn, G., 2008. Über Effizienz hinaus. In: S. Hartard, A. Schaffer, and J. Giegrich, eds. Ressourceneffizienz im Kontext der Nachhaltigkeitsdebatte. Baden-Baden: Nomos.

Schmidt-Bleek, F., 1998. Das MIPS-Konzept. Weniger Naturverbrauch—mehr Lebensqualität durch Faktor 10. München: Droemer Knaur.

Schmidt, M.G., 2004. Wörterbuch zur Politik. Stuttgart: A. Kröner Verlag.

Schmidtz, D., 2004. Satisficing as a Humanly Rational Strategy. In: M. Byron, ed. Satisficing and Maximizing: Moral Theorists on Practical Reason. Cambridge (etc.): Cambridge University Press, 30–58.

Schmitt, C., 1991 [1932]. Der Begriff des Politischen. Berlin: Duncker & Humblot.

Schneidewind, U. and Palzkill-Vorbeck, A., 2011. Suffizienz als Business Case: Nachhaltiges Ressourcenmanagement als Gegenstand einer transdisziplinären Betriebswirtschaftslehre. Wuppertal.

Schneidewind, U. and Zahrnt, A., 2014. The Politics of Sufficiency: Making it easier to live the Good Life. München: oekom Verlag.

Schor, J.B., 2005 a. Prices and quantities: Unsustainable consumption and the global economy. Ecological Economics, 55 (3), 309–320.

Schor, J.B., 2005 b. Sustainable Consumption and Worktime Reduction. Journal of Industrial Ecology, 9 (1-2), 37–50.

Schor, J.B., 2010. Plenitude—The New Economics of True Wealth. New York: Penguin Press.

Schrader, U., 2011. Verbraucherrechte und Verbraucherverantwortung für nachhaltigen Konsum. In: L. Heidbrink, I. Schmidt, and B. Ahaus, eds. Die Verantwortung des Konsumenten. Über das Verhältnis von Markt, Moral und Konsum. Frankfurt, New York: Campus Verlag, 75–92.

Schramme, T., 1999. Verteilungsgerechtigkeit ohne Verteilungsgleichheit. Analyse & Kritik, 21, 171–191.

Schramme, T., 2003. Die Anmaßung der Gleichheitsvoraussetzung. Deutsche Zeitschrift für Philosophie, 51 (2), 255–273.

Schramme, T., 2004. Zur Währungskrise der egalitären Gerechtigkeit (Teil 2). Zeitschrift für philosophische Forschung, 58 (1), 129–141.

Schramme, T., 2006. Is Rawlsian Justice Bad for the Environment? Analyse & Kritik, 28, 146–157.

Schramme, T., 2011. When consumers make environmentally unfriendly choices. Environmental Politics, 20 (3), 340–355.

Schultz, J., Brand, F., Kopfmüller, J., and Ott, K., 2008. Building a 'theory of sustainable development': two salient conceptions within the German discourse. International Journal for Environment and Sustainable Development, 7 (4), 465–482.

Schumacher, E.F., 1973. Small is Beautiful: Economics as if People Mattered. London: Blond & Briggs.

Schwenkenbecher, A., 2013. Joint Duties and Global Moral Obligations. Ratio, 26 (3), 310–328.

Seidl, I. and Zahrnt, A., 2012. Suffizienzpolitik—Damit einfacher leben einfacher wird. Umwelt aktuell, 2–3.

Seidl, I. and Zahrnt, A., eds., 2010. Postwachstumsgesellschaft. Konzepte für die Zukunft. Marburg: Metropolis.

Sellke, P. and Renn, O., 2010. Risk, Society and Environmental Policy: Risk Governance in a Complex World. In: M. Groß and H. Heinrichs, eds. Environmental Sociology: European Perspectives and Interdisciplinary Challenges. New York (u.a.): Springer, 295–321.

Sen, A., 1979. Equality of What? The Tanner Lectures on Human Values. http://tannerlectures.utah.edu/_documents/a-to-z/s/sen80.pdf (last accessed August 2013).

Sen, A., 1992. Inequality Reexamined. Cambridge (Mass.): Harvard University Press.

Sen, A.K., 1999. Democracy as a Universal Value. Journal of Democracy, 10 (3), 3–17.

Seyfang, G., Lorenzoni, I., and Nye, M., 2009. Personal Carbon Trading: a critical examination of proposals for the UK.

Shaheen, S.A. and Cohen, A.P., 2013. Carsharing and Personal Vehicle Services: Worldwide Market Developments and Emerging Trends. International Journal of Sustainable Transportation, 7 (1), 5–34.

Shove, E. and Warde, A., 1998. Inconspicuous consumption: the sociology of consumption and the environment. http://www.lancs.ac.uk/fass/sociology/papers/shove-warde-inconspicuous-consumption.pdf (last accessed June 2014).

Shove, E., 2003. Converging Conventions of Comfort, Cleanliness and Convenience. Journal of Consumer Policy, 26, 395–418.

Shrader-Frechette, K., 1991. Risk and Rationality—Philosophical Foundations for Populist Reforms. Berkeley (etc.): University of California Press.

Shrader-Frechette, K., 1998. Risk. In: E. Craig, ed. Routledge Encyclopedia of Philosophy. London, New York: Routledge, 331–334.

Shue, H., 1993. Substistence Emissions and Luxury Emissions. Law & Policy, 15 (1), 39–59.

Shue, H., 1999. Global Environment and International Inequality. International Affairs, 75 (3), 531–545.

Siebenhüner, B., 2011. Kann die Politik es richten? Konsument(inn)en als politische Akteure. GAIA, 20 (1), 14–16.

Simon, H.A., 1955. A Behavioural Model of Rational Choice. Quarterly Journal of Economics, 69 (1), 99–118.

Simon, H.A., 1965. Administrative Behaviour: A Study of Decision-Making Process in Administrative Organization. New York: The Free Press.

Simon, J.L., 1998. Scarcity or Abundance? In: L. Westra and P.H. Werhane, eds. The Business of Consumption: Environmental Ethics and the Global Economy. Lanham (etc.): Rowman and Littlefield, 237–245.

Singer, P., 2004. Outsiders: our obligations to those beyond our borders. In: D.K. Chatterjee, ed. The ethics of assistance: Morality and the distant needy. Cambridge (etc.): Cambridge University Press, 11–32.

Sinnott-Armstrong, W., 2010. It's Not My Fault. In: S.M. Gardiner, S. Caney, D. Jamieson, and H. Shue, eds. Climate Ethics—Essential Readings. Oxford (etc.): Oxford University Press, 332–344.

Skidelsky, R. and Skidelsky, E., 2012. How much is enough? Money and the good life. New York: Other Press.

Slote, M., 2004. Two views of Satisficing. In: M. Byron, ed. Satisficing and Maximizing: Moral Theorists on Practical Reason. Cambridge (etc.): Cambridge University Press, 14–29.

Sloterdijk, P., 2010. How big is 'big'? http://www.collegium-international.org/index.php/en/contributions/127-how-big-is-big (last accessed September 2014).

Smeddinck, U., 2011. Regulieren durch 'Anstoßen'—Nachhaltiger Konsum durch gemeinwohlverträgliche Gestaltung von Entscheidungssituationen? Die Verwaltung, 44 (3), 375–395.

Smiley, M., 2011. Collective Responsibility. The Stanford Encyclopedia of Philosophy (Fall 2011 Edition). http://plato.stanford.edu/archives/fall2011/entries/collective-responsibility/ (last accessed January 2013).

Smith, A., 2005 [1776]. An Inquiry into the Nature and Causes of the Wealth of Nations.

Socolow, R.H., 1994. Six Perspectives from Industrial Ecology. In: R.H. Socolow, C. Andrews, F. Berkhout, and V. Thomas, eds. Industrial Ecology and Global Change. Cambridge (etc.): Cambridge University Press.

Solow, R.M., 1974. The Economics of Resources or the Resources of Economics. American Economic Review, Papers and Proceedings of the Eighty-sixth Annual Meeting of the American Economic Association (May, 1974), 64 (2), 1–14.

Soper, K. and Trentmann, F., eds., 2008. Citizenship and Consumption. Basingstoke, Hampshire (etc.): Palgrave Macmillan.

Soper, K., 2006. Conceptualizing needs in the context of consumer politics. Journal of Consumer Policy, 29 (4), 355–372.

Soper, K., 2007. Re-thinking the 'Good Life'—The citizenship dimension of consumer disaffection with consumerism. Journal of Consumer Culture, 7 (2), 205–229.

Soper, K., 2008 a. Alternative Hedonism, Cultural Theory and the Role of Aesthetic Revisioning. Cultural Studies, 22 (5), 567–587.

Soper, K., 2008 b. 'Alternative Hedonism' and the Citizen-Consumer. In: K. Soper and F. Trentmann, eds. Citizenship and Consumption. Basingstoke, Hampshire (etc.): Palgrave Macmillan, 191–205.

Spaargaren, G. and Cohen, M.J., 2009. Greening lifecycles and lifestyles: Sociotechnical innovations in consumption and production as core concerns of ecological modernisation theory. In: A.P.J. Mol, D.A. Sonnenfeld, and G. Spaargaren, eds. The ecological modernisation reader: environmental reform in theory and practice. London, New York: Routledge, 257–274.

Spaargaren, G. and Vliet, B. van, 2000. Lifestyles, Consumption and the Environment: The Ecological Modernisation of Domestic Consumption. In: A.P.J. Mol and D.A. Sonnenfeld, eds. Ecological Modernisation around the world: perspectives and critical debates. London: Frank Cass Publishers, 50–76.

Spaargaren, G., 2000. Ecological modernization theory and domestic consumption. Journal of Environmental Policy & Planning, 2 (4), 323–335.

Spaargaren, G., 2009. Sustainable consumption: A theoretical and environmental policy perspective. In: A.P.J. Mol, D.A. Sonnenfeld, and G. Spaargaren, eds. The ecological modernisation reader: environmental reform in theory and practice. London, New York: Routledge, 318–333.

Spaemann, R., 2007. Grenzen der Verantwortung. In: L. Heidbrink and A. Hirsch, eds. Staat ohne Verantwortung? Zum Wandel der Aufgaben von Staat und Politik. Frankfurt, New York: Campus Verlag, 37–53.

Spector, H., 2010. Four Conceptions of Freedom. Political Theory, 38 (6), 780–808.

Spengler, L., 2013. Suffizienzpolitik—Eine heikle Mission. Politische Ökologie, (134), 45–50.

Spengler, L., 2016 a. Two types of 'enough': Sufficiency as minimum and maximum. Environmental Politics, 25 (5), 921–940.

Spengler, L., 2016 b. Ökodesignanforderungen—von der Energie- zur Ressourceneffizienz? In: F. Reimer, ed. Ressourceneffizienz—Leitbild für das Umweltrecht? Baden-Baden: Nomos, 229–241.

Spengler, L. and Dick, D., 2007. Selbstverpflichtungen der Wirtschaft: Anmerkungen zu einem überschätzten Instrument. UfU Rundbriefe, 62 (1/2007), 20–25.

Spiegel Online, 2013. Heftiger Widerstand gegen 'Veggie Day' der Grünen, 5 Aug. http://www.spiegel.de/politik/deutschland/fleischloser-tag-widerstand-gegen-veggie-day-der-gruenen-a-914949.html (last accessed August 2013).

Spitznagel, E., 2007. Wirtschaftswachstum als Beschäftigungsgarant? In: S. Rudolph, ed. Wachstum, Wachstum über alles? Ein ökonomisches Leitbild auf dem Prüfstand von Umwelt und Gerechtigkeit. Marburg: Metropolis, 53–73.

SRU (Sachverständigenrat für Umweltfragen), 2002. Für eine neue Vorreiterrolle—Umweltgutachten 2002 des Rates von Sachverständigen für Umweltfragen. Berlin, Bundestags-Drucksache 14/8792.

SRU (Sachverständigenrat für Umweltfragen), 2012. Umweltgutachten 2012—Verantwortung in einer begrenzten Welt. Berlin: Erich Schmidt Verlag.

Starkey, R., 2012 a. Personal carbon trading: A critical survey. Part 1: Equity. Ecological Economics, 73, 7–18.

Starkey, R., 2012 b. Personal carbon trading: A critical survey. Part 2: Efficiency and effectiveness. Ecological Economics, 73, 19–28.

Steen-Olsen, K., Weinzettel, J., Cranston, G., Ercin, A.E., and Hertwich, E.G., 2012. Carbon, Land, and Water Footprint Accounts for the European Union: Consumption, Production, and Displacements through International Trade. Environmental Science & Technology, 46, 10883–10891.

Steffen, W., Richardson, K., Rockström, J., Cornell, S.E., Fetzer, I., Bennett, E.M., Biggs, R., Carpenter, S.R., de Vries, W., de Wit, C.A., Folke, C., Gerten, D., Heinke, J., Mace, G.M., Persson, L.M., Ramanathan, V., Reyers, B., and Sörlin, S., 2015. Planetary boundaries: Guiding human development on a changing planet. Science, 347 (6223).

Steffen, W., Sanderson, A., Tyson, P., Jäger, J., Matson, P., Moore III, B., Oldfield, F., Richardson, K., Schellnhuber, H.J., Turner, B.L.I., and Wasson, R.J., 2004. Global Change and the Earth System: A Planet Under Pressure. Berlin (etc.): Springer.

Steinberger, J.K., Krausmann, F., Getzner, M., Schandl, H., and West, J., 2013. Development and Dematerialization: An International Study. PLoS ONE, 8 (10), e70385.

Stengel, O., 2011. Suffizienz—die Konsumgesellschaft in der ökologischen Krise. München: oekom Verlag.

Stephens, P.H.G., 2001 a. Green Liberalisms: Nature, Agency and the Good. Environmental Politics, 10 (3), 1–22.

Stephens, P.H.G., 2001 b. The Green only Blooms amid the Millian Flowers. A Reply to Marcel Wissenburg. Environmental Politics, 10 (3), 43–47.

Stern, 2013. FDP-Abgeordneter reizt Grüne mit Nazi-Plakat, 6 Aug. http://www.stern.de/politik/deutschland/kritik-am-veggie-day-fdp-abgeordneter-reizt-gruene-mit-nazi-plakat-3368892.html (last accessed August 2013).

Stern, P.C., 1997. Toward a Working Definition of Consumption for Environmental Research and Policy. In: P.C. Stern, T. Dietz, V.W. Ruttan, R.H. Socolow, and J.L. Sweeney, eds. Environmentally Significant Consumption: Research Directions. Washington, DC: National Academy Press, 12–25.

Stiglitz, J.E., Sen, A., and Fitoussi, J.-P., 2009. Report by the Commission on the Measurement of Economic Performance and Social Progress.

Stokes, S., 1998. Pathologies of Deliberation. In: J. Elster, ed. Deliberative Democracy. Cambridge (u.a.): Cambridge University Press, 123–139.

Sunstein, C.R. and Reisch, L.A., 2014. Automatically Green: Behavioural Economics and Environmental Protection. Harvard Environmental Law Review, 38, 127–158.

Sunstein, C.R., 2005. Laws of Fear: Beyond the Precautionary Principle. Cambridge (etc.): Cambridge University Press.

Sustainable Development Commission, 2009. A Sustainable New Deal—A stimulus package for economic, social and ecological recovery. London.

Talisse, R.B., 2005. Democracy after Liberalism. Pragmatism and Deliberative Politics. New York, London: Routledge.

Temkin, L.S., 2003 a. Equality, Priority or What? Economics and Philosophy, 19 (1), 61–87.

Temkin, L.S., 2003 b. Egalitarianism Defended. Ethics, 113, 764–782.

Tertoolen, G., van Kreveld, D., and Verstraten, B., 1998. Psychological Resistance Against Attempts to Reduce Private Car Use. Transportation Research Part A, 32 (3), 171–181.

Thaler, R.H. and Sunstein, C.R., 2008. Nudge: Improving Decisions About Health, Wealth, and Happiness. New Haven, Conn.: Yale University Press.

Thoreau, H.D., 2010 [1854]. Walden. Princeton: Princeton University Press.

Tremmel, J., 2008. A Theory of Intergenerational Justice. Düsseldorf.

Tribe, L.H., 1980. Was spricht gegen Plastikbäume? In: D. Birnbacher, ed. Ökologie und Ethik. Stuttgart: Reclam, 20–71.

Tukker, A., Cohen, M.J., Hubacek, K., and Mont, O., 2010. The Impacts of Household Consumption and Options for Change. Journal of Industrial Ecology, 14 (1), 13–30.

Tukker, A., Eder, P., and Suh, S., 2006. Environmental Impacts of Products: Policy Relevant Information and Data Challenges. Journal of Industrial Ecology, 10 (3), 183–198.

Turner, G.M., 2014. Is Global Collapse Imminent? Melbourne Sustainable Society Institute, The University of Melbourne.

UBA (Umweltbundesamt), ed., 2012. Glossar zum Ressourcenschutz. Dessau-Roßlau.

UBA (Umweltbundesamt), ed., 2014 a. Soziale Innovationen im Aufwind: Ein Leitfaden zur Förderung sozialer Innovationen für nachhaltigen Konsum. Dessau-Roßlau.

UBA (Umweltbundesamt), ed., 2014 b. Klimaneutral leben: Verbraucher starten durch beim Klimaschutz. Dessau-Roßlau.

Ullmann-Margalit, E., 1976. The generalization argument: Where does the obligation lie? Journal of Philosophy, 73 (15), 511–522.

UN (United Nations), 1992. Convention on Biological Diversity.

UN (United Nations), 2015. Transforming Our World: The 2030 Agenda for Sustainable Development. No. A/RES/70/1.

UN (United Nations), 2015 a. The Millennium Development Goals Report. New York.

UNCED (United Nations Conference on Environment & Development), 1992 a. Agenda 21. Rio de Janeiro.

UNCED (United Nations Conference on Environment & Development), 1992 b. Rio Declaration on Environment and Development.

UNCTAD (United Nations Commission on Trade and Development) and UNEP (United Nations Environment Programme), 1974. Cocoyoc Declaration. Cocoyoc.

UNEP (United Nations Environment Programme), 2009. Global Green New Deal Policy Brief.

UNEP (United Nations Environment Programme), 2011 a. Decoupling natural resource use and environmental impacts from economic growth. A Report of the Working Group on Decoupling to the International Resource Panel. Nairobi.

UNEP (United Nations Environment Programme), 2011 b. Towards a Green Economy: Pathways to Sustainable Development and Poverty Eradication.

UNEP (United Nations Environment Programme), 2012. GEO 5—Global Environmental Outlook. Environment for the future we want. Nairobi.

UNEP (United Nations Environment Programme), 2014 a. Decoupling 2: Technologies, Opportunities and Policy Options. A Report of the Working Group on Decoupling to the International Resource Panel.

UNEP (United Nations Environment Programme), 2014 b. Assessing Global Land Use: Balancing Consumption with Sustainable Supply. A Report of the Working Group on Land and Soils of the International Resource Panel.

UNFCCC (United Nations Framework Convention on Climate Change), 2015. Adoption of the Paris Agreement; Conference of the Parties, Twenty-first session. Paris, No. FCCC/CP/2015/L.9/Rev. 1.

Unseld, G., 1991. 'Für oder gegen Technik?' Die Streitfrage um unsere Zukunft. Scheidewege, 21, 59–82.

UNWSSD (United Nations World Summit on Sustainable Development), 2002 a. Johannesburg Declaration on Sustainable Development.

UNWSSD (United Nations World Summit on Sustainable Development), 2002 b. Plan of Implementation of the World Summit on Sustainable Development.

van den Bergh, J.C.J.M., 2011. Environment versus growth—A criticism of 'degrowth' and a plea for 'a-growth'. Ecological Economics, 70 (5), 881–890.

Vanner, R., 2006. Ex-post estimates of costs to business of EU environmental policies: A case study looking at Ozone Depleting Substances. European Commission.

Veblen, T., 1961 [1899]. The theory of the leisure class: An economic study of institutions. New York: Random House.

Verstreng, V., Myhre, G., Fagerli, H., Reis, S., and Tarrasón, L., 2007. Twenty-five years of continuous sulphur dioxide emission reduction in Europe. Atmospheric Chemistry and Physics, 7, 3663–3681.

Vincent, A., 1998. Liberalism and the environment. Environmental Values, 7 (4), 443–459.

Voget-Kleschin, L., 2012. Sustainable Food Consumption? Claims for Sustainable Lifestyles in between Normative and Eudaimonistic Issues: The Example of Food Production and Consumption.

Voget, L., 2009. Suffizienz als politische Frage. In: T. v. Egan-Krieger, J. Schultz, P.P. Thapa, and L. Voget, eds. Die Greifswalder Theorie starker Nachhaltigkeit. Ausbau, Anwendung und Kritik. Marburg: Metropolis, 209–224.

Waldron, J., 1989. Legislation and Moral Neutrality. In: R.E. Goodin and A. Reeve, eds. Liberal Neutrality. London, New York: Routledge, 61–83.

Wall, S. and Klosko, G., eds., 2003. Perfectionism and Neutrality: Essays in Liberal Theory. Lanham (etc.): Rowman and Littlefield.

Wall, S., 2003. Just Savings and the Difference Principle. Philosophical Studies, 116, 79–102.

Wall, S., 2012. Perfectionism in Moral and Political Philosophy. The Stanford Encyclopedia of Philosophy (Winter 2012 Edition). http://plato.stanford.edu/archives/win2012/entries/perfectionism-moral/ (last accessed June 2013).

Ware, A. and Goodin, R.E., eds., 1990. Needs and Welfare. London (etc.): Sage Publications.

Waschkuhn, A., 1998. Demokratietheorien—Politiktheoretische und ideengeschichtliche Grundzüge. München, Wien: R. Oldenbourg Verlag.

WBGU (Wissenschaftlicher Beirat der Bundesregierung Globale Umweltveränderungen), 1998. Strategien zur Bewältigung Globaler Umweltrisiken. Jahresgutachten 1998. Berlin (etc.): Springer.

WCED (World Commission on Environment and Development), 1987. Our Common Future. Report of the World Commission on Environment and Development. Oslo.

Weaver, P., Jansen, L., Van Grootveld, G., Van Spiegel, E., and Vergragt, P., 2000. Sustainable Technology Development. Sheffield: Greenleaf Publishing.

Weber, M., 1992 [1919]. Politik als Beruf. Stuttgart: Reclam.

Weinstock, D.M., 2003. Miller on Distributive Justice. In: D.A. Bell and A. de-Shalit, eds. Forms of Justice: Critical Perspectives on David Miller's Political Philosophy. Lanham: Rowman and Littlefield, 269–286.

Weizsäcker, E.U. v., Hargroves, K., Smith, M., Deisha, C., and Stasinopoulos, P., 2009. Factor Five. Transforming the Global Economy through 80% Improvements in Resource Productivity. London: Earthscan.

Weizsäcker, E.U. v., Lovins, A.B., and Lovins, L.H., 1995. Faktor Vier: Doppelter Wohlstand—halbierter Naturverbrauch. München: Droemer Knaur.

Wells, P. and Zapata, C., 2012. Renewable Eco-industrial Development: A New Frontier for Industrial Ecology? Journal of Industrial Ecology, 16 (5), 665–668.

Wenar, L., 2011. Rights. The Stanford Encyclopedia of Philosophy (Fall 2011 Edition). http://plato.stanford.edu/archives/fall2011/entries/rights/ (last accessed October 2014).

Westra, L. and Werhane, P.H., eds., 1998. The Business of Consumption: Environmental Ethics and the Global Economy. Lanham (etc.): Rowman and Littlefield.

White, D.F., 2002. A Green Industrial Revolution? Sustainable Technological Innovation in a Global Age. Environmental Politics, 11 (2), 1–26.

White, S., 2008. Social Minimum. The Stanford Encyclopedia of Philosophy (Fall 2008 Edition). http://plato.stanford.edu/archives/fall2008/entries/social-minimum/ (last accessed September 2014).

Wiebe, K.S., Bruckner, M., Giljum, S., Lutz, C., and Polzin, C., 2012. Carbon and Materials Embodied in the International Trade of Emerging Economies: A Multiregional Input-Output Assessment of Trends Between 1995 and 2005. Journal of Industrial Ecology, 16 (4), 636–646.

Wieland, W., 1999. Verantwortung—Prinzip der Ethik? Vorgetragen am 28. Juni 1997. Heidelberg: Universitätsverlag C. Winter.

Winterfeld, U. v., 1993. Über die Kunst des richtigen Verhaltens. Politische Ökologie, (Special Sept./Okt. '93), 45–47.

Winterfeld, U. v., 2002. Reflexionen zur Suffizienz als politischer Angelegenheit in sieben Etappen. In: M. Linz, P. Bartelmus, P. Hennicke, R. Jungkeit, W. Sachs, G. Scherhorn, G. Wilke, and U. v. Winterfeld, eds. Von nichts zu viel—Suffizienz gehört zur Zukunftsfähigkeit. 27–37.

Winterfeld, U. v., 2007. Keine Nachhaltigkeit ohne Suffizienz. Fünf Thesen und Folgerungen. vorgänge, (3), 46–54.

Wissenburg, M., 1998. Green liberalism. The free and the green society. London (etc.): UCL Press.

Wissenburg, M., 2001 a. Liberalism is Always Green on the Other Side of the Mill. A Reply to Piers Stephens. Environmental Politics, 10 (3), 23–42.

Wissenburg, M., 2001 b. Sustainability and the Limits of Liberalism. In: J. Barry and M. Wissenburg, eds. Sustaining liberal democracy. Ecological challenges and opportunities. New York (etc.): Palgrave, 192–204.

Wissenburg, M., 2006 a. Ecological Neutrality and Liberal Survivalism—How (not) to Discuss the Compatibility of Liberalism and Ecologism. Analyse & Kritik, 28, 125–145.

Wissenburg, M., 2006 b. Liberalism. In: A. Dobson and R. Eckersley, eds. Political Theory and the Ecological Challenge. Cambridge, New York: Cambridge University Press, 20–34.

Withers, H., 1920. The Consumer's Responsibility. Annals of the American Academy of Political and Social Science, 87, 225–232.

WWF, 2014 a. Living Planet Report 2014: Species and spaces, people and places. Gland, Switzerland.

WWF, 2014 b. Fleisch frisst Land. Berlin.

Wynne, B., 1992. Uncertainty and environmental learning: Reconceiving science and policy in the preventive paradigm. Global Environmental Change, 2 (2), 111–127.

Xenos, N., 1987. Liberalism and the Postulate of Scarcity. Political Theory, 15 (2), 225–243.

Young, I.M., 2004. Responsibility and Global Labor Justice. The Journal of Political Philosophy, 12 (4), 365–388.

Young, I.M., 2006. Responsibility and Global Justice: A Social Connection Model. Social Philosophy and Policy, 23 (1), 102–130.

Young, I.M., 2011. Responsibility for Justice. Oxford, New York: Oxford University Press.

Zapf, W., 1989. Über soziale Innovationen. Soziale Welt, 40 (1/2), 170–183.